$$VR \frac{(14.7)\left(\overset{Time}{6\,min}\right)(20cfm - 0)}{max - min} \times 7.98$$

8

$$P_2 = \frac{P_1 V_1 T_2}{V_2 T_1}$$

Hydraulics for Engineering Technology

James E. Johnson, P.E.

Prentice Hall
Englewood Cliffs, New Jersey   Columbus, Ohio

Library of Congress Cataloging-in-Publication Data

Johnson, James E.
    Hydraulics for engineering technology / James E. Johnson.
        p.   cm.
    Includes index.
    ISBN 0-13-232513-6
    1. Oil hydraulic machinery.   2. Hydraulic control.   3. Hydraulics.
    I. Title.
    TJ843.J48 1996
    621.2--dc20   95-43128
                                                            CIP

Editor: Stephen Helba
Production Editor: Rex Davidson
Cover Designer: Proof Positive/Farrowlyne Assoc., Inc.
Production Buyer: Laura Messerly
Marketing Manager: Debbie Yarnell
Illustrations: Diphrent Strokes, Inc.

This book was set in Times Roman by The Clarinda Company and was printed and bound by
Quebecor Printing/Book Press. The cover was printed by Phoenix Color Corp.

 © 1996 by Prentice-Hall, Inc.
A Simon & Schuster Company
Englewood Cliffs, New Jersey 07632

Printed in the United States of America

10 9 8 7 6 5 4 3 2 1

ISBN: 0-13-232513-6

Prentice-Hall International (UK) Limited, *London*
Prentice-Hall of Australia Pty. Limited, *Sydney*
Prentice-Hall Canada Inc., *Toronto*
Prentice-Hall Hispanoamericana, S. A., *Mexico*
Prentice-Hall of India Private Limited, *New Delhi*
Prentice-Hall of Japan, Inc., *Tokyo*
Simon & Schuster Asia Pte. Ltd., *Singapore*
Editora Prentice-Hall do Brasil, Ltda., *Rio de Janeiro*

Preface

Most practical hydraulic systems use electrical or electronic components to control or power the system. Because of this, many companies involved in the development, sales, and servicing of these systems require that their engineers and technicians be skilled in electrotechnology as well as hydraulics.

Compatibility of Textbooks

Hydraulics and electrotechnology may be learned together or separately. In either case, however, the hydraulics and electric textbooks should be compatible with. One good way of learning these two fields is to study hydraulics as an analog of electronics and electronics as an analog of hydraulics.

Not "So Simple That It's Wrong"

This is not a physics book, nor is it an electrical text; yet it does not violate laws of physics or electronics. It is not "so simple that it's wrong." It is a basic hydraulics text from which students can move into electronics or advance to higher levels of hydraulics or other engineering subjects without having to unlearn basics.

Concepts and Theory as Needed

Students are not burdened with all theory and concepts being presented in the first part of this book. Concepts and theory are presented when needed, as in good teaching practice.

Your Electrical Knowledge Will Help

This book does not attempt to teach hydraulics and electrotechnology at the same time. But it presents the material in such a way that you may use your electrical knowledge to better understand hydraulics and vice versa. Some electrohydraulic analogies are drawn, and the path is left open for a two-way crossover between electrical and hydraulic fundamentals.

Dimensional Calculations

Too many hydraulics students spend far too much time memorizing formulas for solving problems by the "rote" method. They do all of this work without ever learning the true meaning of the problems or knowing whether the answers are correct. This book teaches dimensional calculations, which allows the student to solve most hydraulic problems without formulas. When the dimensions of the answer fall into their proper places, the student will have the correct answer.

Although formulas are not needed in the solution of most dimensional problems, formulas are generally shown in this text when introducing new calculations. This is done to ease the transition from formula to dimensional calculations. But, by all means, the student should be encouraged to use dimensions and avoid using the formula as a crutch. My hope is that before the student is finished with this book he or she can leave the "little black book of formulas" at home.

U.S. and English Units versus SI Units

The U.S. units used in this text are the same as those of the English system of units except that the U.S. gallon (231 in.^{3}) and short ton (2000 lb) are used instead of the imperial gallon (277.42 in.^{3}) and long ton (2240 lb).

Système International (SI) units are covered in Chapter 2 and are used in many chapters throughout the book. With dimensional calculations, problems may be solved using U.S. units, English units, SI units, or any mixture of the three.

At the beginning of the book, a complete table of contents is provided for each chapter. All topics and subheadings are called out, along with examples, figures, and tables. This comprehensive table of contents for each chapter provides a rather helpful quick reference for use after graduation or for easy review during the course.

Page Layout

The layout of the book is designed so that the illustrations appear as close to their reference point and text discussion. Due to the large size of some arts this ideal layout is not always possible. Readers may find it helpful to photocopy those arts not falling immediately near their references to keep them at hand while reading the text discussion.

Review Summary

Pertinent summary statements are included—and numbered for easy reference—at the end of each chapter. These summaries are handy for the teacher as test material and for the student as review.

Problems and Questions

Problems and questions are available at the end of each chapter. These may be used for study, review, or quizzes. Solutions to odd-numbered problems are in Appendix C. The instructor may request that these problems be recalculated using slightly different dimensions.

A glossary of terms (Appendix A), a chart of graphical symbols used in hydraulic circuits (Appendix B), and an index are added to cover adequately the material of this book.

This book will be useful to (1) those studying for an associate degree in engineering technology, (2) those in high school or post–high school technicians programs, (3) those studying for a B.S. degree in engineering technology, (4) those already working in the field of hydraulics or electrohydraulics who want a different slant on the subject, and (5) those in other technologies who can benefit from dimensional calculations.

The book is not only useful to students but also to the practitioners and novices who are already engaged in the field.

Acknowledgments

Any knowledge of hydraulics that the author has acquired over the years was not self-generated. Many people made important contributions—writers, teachers, colleagues, customers, students, and other fellow-workers. Some contributed directly to the book, but countless others made their contributions through discussions, lectures, and writings. While the listing of all contributors would be impossible, I would like to include some outstanding examples that come to mind:

Ed Brown, Bob Thoren, Jack Walrad, and Royce Saari of Vickers Inc. for their help and encouragement.

John Marshal of Vickers Inc. for his encouragement and help with the text, especially SI standards.

George Altland of Vickers Inc., who taught me the fundamentals of hydraulics.

George McConnell of Vickers Inc., who instilled in me (mostly by example) the advantages of using self-checking dimensional analysis.

Bill Bonham of Vickers Inc. for his encouragement and our brain-storming session, which resulted in the design of the level detector for the NASA Crawler Transporter.

Paul Noble of Fellows Gear Shaper, Vt. for his idea of the double-flow diagram of energy transfer, such as shown in Figures 7–16 and 9–1.

Ray Spencer of Vickers Inc. for his help and paper on the "Advantages of Hydraulic Servo".

I wish to thank the following reviewers of the manuscript for their helpful comments and suggestions:

Montie Fleshman, Wytheville Community College; John Marshall; George A. McConnell; Dick Minch, Waukesha County Technical College; Charles Sorenson, Ivy Tech; Ken Warfield, Shawnee State University.

Brief Contents

Contents

1
Principles of Hydraulics   1

Figures

2
Dimensional Calculations 15

Tables

3
Hydraulic Actuators   37

Examples

Figures

4
Directional Control 59

Examples

Figures

5
Pressure Sensing and Control   79

Figures

6
Flow Control 101

7
Hydraulic Pumps 127

8
Nonservo Circuits and Systems

151

Example

Figures

Tables

9
Electrohydraulic Servo Systems

187

Figures

Tables

10
Troubleshooting Guide and Maintenance Hints   249

1
Principles of Hydraulics

The history of hydraulics goes back to ancient times. Depictions of pumps and water-wheels are seen in some of the earliest recorded history. But in the seventeenth century a French scientist, Blaise Pascal, discovered the principles of hydraulics with which we are concerned here. Pascal's law has to do with confined fluids used to transmit power, multiply forces, and cause motion.

The story of hydraulics is mostly about pressure and flow. Pressure, which is force per unit area, can be measured in U.S. units as $lb/in.^2$ [pounds per square inch (sometimes called psi)], and in SI, or metric, units as N/m^2 [newtons per square meter (sometimes called pascals, Pa)], or in bars where 1 bar = 14.5 $lb/in.^2$ or 100 kPa. Flow can be measured in U.S. units as $in.^3/min$ (cubic inches per minute) or gal/min (gallons per minute), and in SI units as cm^3/min (cubic centimeters per minute) or L/min (liters per minute).

1.1 HEAD PRESSURE

The first pressure of which we should be aware, even before the pump is started, is the pressure caused by the weight of the fluid in the system, or the fluid density [weight per unit volume $(lb/in.^3)$] and height of the fluid (ft) above the test point. In this book we call this the *head pressure*.

Figure 1.1 shows three separate containers of fluids depicting the effects of fluid height, volume, and container shape on head pressure.

Effect of Fluid Height on Head Pressure

Figure 1.1a shows a square container with 1 ft^3 of oil (1 ft by 1 ft by 1 ft). The weight of the oil is approximately 58 lb. Since the area of the tank bottom is 12 in. × 12 in. = 144 $in.^2$, the weight of the oil above each square inch of the tank bottom is 58 lb/144 = 0.4 lb. Therefore, the head pressure is 58 lb/144 $in.^2$ = 0.4 $lb/in.^2$. If the oil level were reduced to a height of 6 in., the weight of the oil would now be 29 lb and the head pressure would be 29 lb/144 $in.^2$ = 0.2 $lb/in.^2$. *So, the head pressure is halved as the height of the fluid is halved.*

1

Effect of Fluid Volume on Head Pressure

Figure 1.1b shows a column of oil 1 ft high in a 1-in. by 1-in. square-sided container. Since the area of the container bottom is 1 in.^{2}, and the weight of the oil is 0.4 lb (58 lb/144 = 0.4 lb), the head pressure at the bottom of the tank is 0.4 lb/in.^{2} (the same pressure as that of the larger tank of Figure 1.1a). *So, the head pressure remains the same (0.4 lb/in.^{2}) when the oil height is held constant at 1 ft even when the volume of the oil is changed. Even if the area of the tank base were 1 acre, the pressure for an oil height of 1 ft would still be 0.4 lb/in.^{2}.*

Effect of Container Shape on Head Pressure

Figure 1.1b shows a square-sided tank 1 ft by 1 ft by 6 in. high. The tank is sealed except for a square-sided tube 1 in. by 1 in. that extends from the top. The tank and tube are

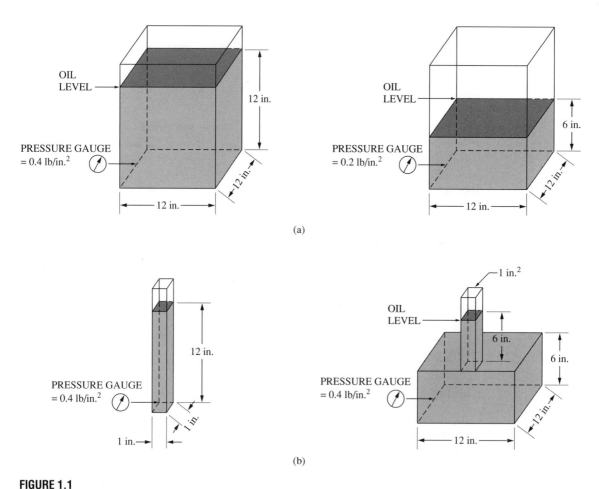

FIGURE 1.1
(a) Head pressure is proportional to fluid height. (b) Head pressure does not change with volume alone (only with height).

filled with oil to a level of 6 in. into the tube. The weight of the oil in the tank is 29 lb and the oil inside the tube weighs 0.2 lb.

The head pressure at the bottom of the tube is 0.2 lb/in.^{2}. The pressure difference from the top of the tank to the bottom (sometimes called psid) is 29 lb/144 in.^{2} = 0.2 lb/in.^{2}. The head pressure at the bottom of the tube acts on the entire area of the fluid at the top of the tank. So, the pressure at the bottom of the tank is the sum of the pressure difference of the tank plus the pressure at the top of the tank or

$$0.2 \text{ lb/in.}^2 + 0.2 \text{ lb/in.}^2 = \textbf{0.4 lb/in.}^2$$

This is the same head pressure that we saw at the bottom of the tanks of both Figures 1.1a and 1.1.b. *So, the head pressure is proportional to the height of the fluid regardless of the shape of the container.*

Significant Head Pressure

Head pressures are always present in hydraulic systems, but frequently they can be ignored as insignificant when dealing with higher system pressures. There are three cases, however, for which head pressures are quite significant: (1) where the other pressures involved are very low, such as at the inlet of a pump; (2) where the height of the fluid is great, such as from the bottom of the ocean; or (3) where the area affected by the head pressure is very large.

Destructive Head Pressure

Figure 1.2 shows a drawing depicting a home basement with a standpipe screwed into the floor drain. This procedure has been tried by home owners in the mistaken belief that it would be a good method of keeping the basement dry should the water in the storm sewer try to back up into the basement. With weeping tile installed under the basement floor for normal drainage, the underside of the floor will be exposed to the backup water, and will be acted on by the head pressure created by the standpipe.

Assume 3 ft of water in the standpipe (measured from the bottom of the floor). For this example, let's assume that the head pressure of water is the same as that of oil:

$$\frac{0.4 \text{ lb/in}^2}{\text{ft}}$$

The head pressure at the bottom of the pipe would be

$$3 \text{ ft} \times \frac{0.4 \text{ lb/in.}^2}{\text{ft}} = \textbf{1.2 lb/in.}^2$$

In other words, with 0.4 lb/in.^{2} of pressure per foot of water, the pressure for 3 ft would be 3×0.4 lb/in.^{2} = 1.2 lb/in.^{2}.

Note: Observe that the unit of measure "ft" has a strikethrough at each location. This means that the "ft" on the top of the fraction cancels the "ft" on the bottom.

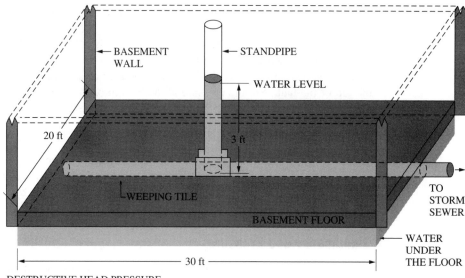

FIGURE 1.2
Basement standpipe.

One might think that 1.2 lb/in.^{2} is insufficient pressure with which to be concerned. But consider the *force* pushing upward from the bottom of the floor. The area of the basement floor is

$$20 \text{ ft} \times \frac{12 \text{ in.}}{\text{ft}} \times 30 \text{ ft} \times \frac{12 \text{ in.}}{\text{ft}} = \textbf{86,400 in.}^2$$

The force pushing the floor upward would be

$$\frac{1.2 \text{ lb}}{\text{in.}^2} \times 86,400 \text{ in.}^2 = \textbf{103,680 lb} \text{ or}$$

$$103,680 \text{ lb} \times \frac{\text{ton}}{2000 \text{ lb}} = \underline{\textbf{51.84 tons}}$$

This force could break the basement floor.

Note: $\left(\dfrac{12 \text{ in.}}{\text{ft}}\right)$ *and* $\left(\dfrac{\text{ton}}{2000 \text{ lb}}\right)$ *are conversion fractions. The numerator (on top) is equal to the denominator (on the bottom), which gives the fraction an absolute value of one. Multiplying by these fractions changes the dimensions but not the absolute value of an equation. See Chapter 2 for further details.*

Of course, the basement would have been flooded had the standpipe not been screwed into the drain. Flooding the basement would have saved the floor, however, because the weight of the floodwater in the basement would have pushed down on the floor with the same magnitude at which the head pressure force was pushing upward, thus preventing damage to the floor.

Use of Head Pressure for Level Control

NASA's Crawler transporter was designed to move a missile and its launch tower from the assembly building to the launch site on Merrit Island. This giant structure is maintained perpendicular during travel by a leveling device, which senses the difference of fluid head pressure at the four corners of the Crawler compared to the tank "bubble" located at the center of the Crawler. More details about this leveling device are given in Chapter 9.

1.2   PASCAL'S LAW

Figure 1.3a shows a jug of wine filled to a height of 1 ft. As usual, the head pressure graduates down the height of the fluid to 0.2 lb/in.^{2} at 6 in. and 0.4 lb/in.^{2} at the bottom of the jug. The area of the jug bottom is 50 in.^{2}, and the force of the bottom caused by the head

FIGURE 1.3
(a) Wine jug head pressure.
(b) Wine jug head pressure with applied force.

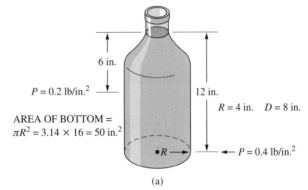

(a)

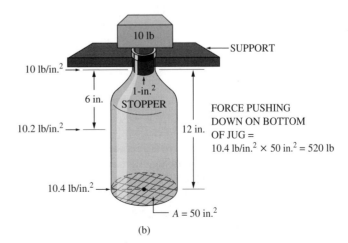

(b)

pressure is 0.4 lb/in.^{2} × 50 in.^{2} = 20 lb. Note that "in.^{2}" has been canceled, in both places, by strikethroughs. The jug will have been designed to carry this weight with, perhaps, a 100% safety margin.

Figure 1.3b shows the same jug with a 1-in.^{2} cork inserted so that it touches the wine (no air pocket). The calculations show what happens if the cork is pushed downward with a force of 10 lb.

Pascal's law states (in effect): *Pressure applied to a confined fluid is transmitted in all directions, and acts with equal force on equal areas, and at right angles to them.* This law, as illustrated at the beginning of this chapter, allows the applied force to be conducted around corners and through irregular passages to the desired destination.

The 10 lb of force per the 1-in.^{2} cork is transmitted to each and every 1 in.^{2} of the jug. So every square inch will see an additional 10 lb of force added to the head pressure force already there. Now the bottom of the jug will see an additional force of

$$\frac{10 \text{ lb}}{\text{in.}^2} \times 50 \text{ in.}^2 = \textbf{500 lb}$$

or a total of 520 lb, counting the head pressure. *This force, which exceeds the safety margins of most jugs, could break the jug.*

Note: The head pressure is generally left out of this type of calculation because the 20 lb of head force is small compared to the applied force transmitted from the cork to the bottom of the jug.

1.3   MECHANICAL AND HYDRAULIC LEVERAGE

Figure 1.4a shows a mechanical weight leverage system where the length of the left arm is 10 times that of the right arm. Any weight placed at point B can be balanced by 1/10 as much weight at point A. To move the large weight up 1 in., the small weight would have to be moved down 10 in. The work needed to move the balanced weights would be very small, only enough to overcome the friction of the pivot.

Figure 1.4b shows a hydraulic leverage system that works the same way. The 10-lb weight placed on the 1-in.^{2} piston at the left transmits a pressure of 10 lb/in.^{2} to the 10-in.^{2} piston on the right. This produces a force of

$$\text{Force} = \text{Pressure} \times \text{Area}$$

$$\frac{10 \text{ lb}}{\text{in.}^2} \times 10 \text{ in.}^2 = \textbf{100 lb}$$

under the large piston, thus supporting the 100-lb weight above the piston and balancing the system.

Another way of describing Figure 1.4b is as follows: The 10-lb weight on the 1-in.^{2} piston produces the same pressure as the 100-lb weight on the 10-in.^{2} piston (both produce 10 lb/in.^{2}). With equal pressures under both pistons, no flow or movement can take place. A change of pressure (ΔP) is required to produce movement of fluid. The system is balanced. The 10-lb weight has balanced the 100-lb weight.

To move the large weight up 1 in., the small weight would have to be moved down 10 in. The work needed to move the balanced weights would be small—only enough to

FIGURE 1.4
(a) Mechanical leverage system.
(b) Hydraulic leverage.

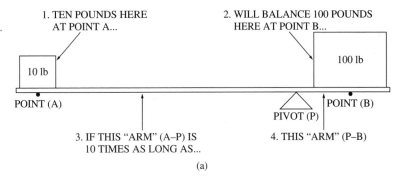

1. TEN POUNDS HERE
AT POINT A...

2. WILL BALANCE 100 POUNDS
HERE AT POINT B...

100 lb

10 lb

POINT (A)

PIVOT (P)

POINT (B)

3. IF THIS "ARM" (A–P) IS
10 TIMES AS LONG AS...

4. THIS "ARM" (P–B)

(a)

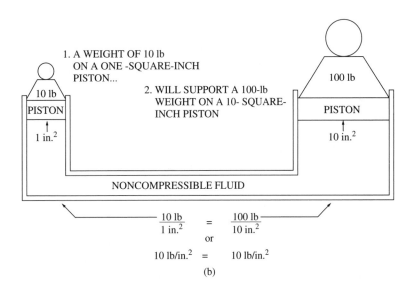

1. A WEIGHT OF 10 lb
ON A ONE-SQUARE-INCH
PISTON...

2. WILL SUPPORT A 100-lb
WEIGHT ON A 10-SQUARE-
INCH PISTON

10 lb

100 lb

PISTON

PISTON

1 in.^{2}

10 in.^{2}

NONCOMPRESSIBLE FLUID

$$\frac{10 \text{ lb}}{1 \text{ in.}^2} = \frac{100 \text{ lb}}{10 \text{ in.}^2}$$

or

$$10 \text{ lb/in.}^2 = 10 \text{ lb/in.}^2$$

(b)

overcome the friction of the two pistons and any viscous drag of the fluid through the pipe, plus a small head pressure caused by the change in fluid heights.

The small piston was pushed down 10 in. while the large piston went up 1 in., leaving an 11-in. difference in liquid levels. This produces a head pressure of almost 0.4 lb/in.^{2} pushing up on the underside of the small piston, requiring an additional 0.4 lb to hold the piston down.

1.4   ENERGY CONSERVATION

A fundamental law of physics states: *Energy can be neither created nor destroyed.* To put it another way, you get out as much energy as you put in, or "energy out equals energy in." Figure 1.5 is similar to Figure 1.4 except that the input weights are replaced by manual forces. This provides a more descriptive example of energy conservation by showing the input as active work rather than static energy.

Figure 1.5a shows an input of 10 lb of force moving the leverage arm down a distance of 10 in., providing an input work (or energy) of 10 in. × 10 lb = 100 in.-lb.

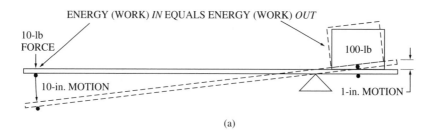

(a)

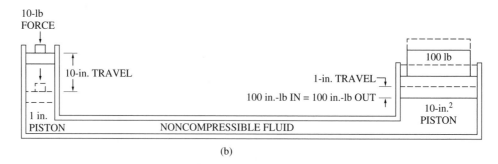

(b)

FIGURE 1.5
Energy (or work) into a mechanical leverage equals energy (or work) out.

Moving the lever arm down 10 in. causes the load weight to move up 1 in., producing an output energy (or work) of 1 in. × 100 lb = **100 in.-lb.** *So, the output work of 100 in.-lb is equal to the input work of 100 in.-lb.*

Figure 1.5b shows an input force of 10 lb applied to a 1-in.² piston, moving it 10 in. This is the work of 10 in. of movement with a force of 10 lb or 100 in.-lb. Since the output cylinder only moved up 1 in., the output energy (or work) is 1 in. times 100 lb or 100 in.-lb.

Note: The distance moved by the small piston must be sufficient to displace a volume of fluid equal to the fluid required to move the larger piston.

The hydraulic jack and hydraulic press both use the principle shown in Figure 1.5b. Small forces are converted to larger forces while the input energy remains equal to the output energy. That is, the product of the input force × distance equals the product of the output force × distance. As a small force moves a long distance, a large force moves a small distance. *This is the law of conservation of energy.*

1.5   THE BASIC PUMP

Hydraulic pumps are energy- or work-converting devices. They convert mechanical work to hydraulic work, just as an electric motor converts electric energy to mechanical. The rate of doing work is called *power.* The output hydraulic horsepower of a pump is equal to its mechanical input horsepower, if you ignore the loss through the pump.

The hydraulic pump does not suck the oil into its inlet, as some people think. The oil must be pushed into the inlet. Then, the pump can push it out.

Positive Inlet Pressure

Figure 1.6 shows a basic pump with a positive inlet pressure. This hand-operated piston pump is connected to a lift cylinder in order to raise a 100-lb weight. Two check valves, V_1 and V_2, are used to direct fluid in the proper direction as the pump is stroked. Each check valve has a ball (or poppet), which is held on its seat by a spring, preventing flow in the wrong direction.

Flow can move in the other direction through the valve (shown by the arrow) when the hydraulic pressure is strong enough to push the ball off its seat. The check valve springs are generally rated by the pressure required to produce flow through the valve. The spring of V_1 requires 1 lb/in.^{2}, and V_2 requires 2 lb/in.^{2} to open and allow flow to take place. The valves of Figure 1.6 are called *angle check valves* because the flow out is at a 90-deg angle from the inlet flow. These valves, with a standard 5 lb/in.^{2} spring, are commonly used in practical hydraulics circuits.

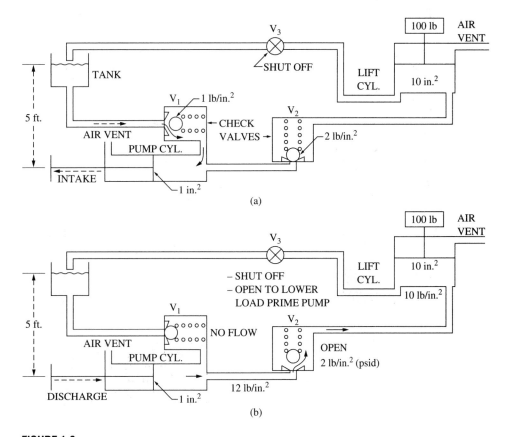

FIGURE 1.6
Basic pump with positive inlet pressure. (a) Intake stroke. (b) Discharge stroke.

How a Positive Inlet Pump Works

The tank of Figure 1.6 is located so that the liquid level is 5 ft higher than the input to the pump. This height produces

$$\frac{0.4 \text{ lb/in.}^2}{\cancel{ft}} \times 5 \cancel{ft} = \textbf{2 lb/in.}^2$$

This is ample pressure to push the ball of V_1 off its seat, with enough left over to push fluid into the pumping chamber as the handle is moved to the left (intake stroke, see Figure 1.6a).

When the pump handle is moved to the right (discharge stroke, see Figure 1.6b), fluid will try to move backwards through V_1. Before this can happen, however, the spring will have seated the ball of V_1.

Now the only way that fluid can get out of the pumping chamber is through V_2 and into the lift cylinder, raising the 100-lb load. The pressure required at the outlet of the pumping chamber to lift the load is

$$\frac{\overset{10}{\cancel{100}} \text{ lb}}{\underset{}{\cancel{10} \text{ in.}^2}} = \textbf{10 lb/in.}^2$$

plus the 2 lb/in.^{2} spring of V_2 for a total of **12 lb/in.^{2}**. The force required to push the pump handle to the right is found by multiplying pressure by the area of the pump piston:

$$\text{Force} = \frac{12 \text{ lb}}{\cancel{\text{in.}^2}} \times 1 \cancel{\text{in.}^2} = \textbf{12 lb}$$

(plus a little to overcome friction and head pressure).

■ | **EXAMPLE 1.1**

Reference Figure 1.6. How far will the piston of the lift cylinder be moved if the pump piston is moved 5 in. to the right?

Solution: The movement of the pistons would be inversely proportional to their areas. The lift piston (10 in.^{2}) would move 1/10 the distance of the pump piston (1 in.^{2}). Or the lift piston would move $1/10 \times 5$ in. = **0.5 in.** ■

Disadvantages of an Elevated Tank

While the positive inlet system of Figure 1.6 has an excellent inlet condition, it also has certain disadvantages:

1. It requires excessive vertical space.
2. Since the tank is higher than the pump, the pump cannot be drained to tank.
3. A leak in the pump circuit could spill fluid from the tank onto the floor.

Negative Inlet Pressure

Figure 1.7 shows a basic pump with the tank mounted below the pumping chamber. This arrangement will solve the problems of Figure 1.6, but, because the tank is below the pumping circuit, the pump inlet will see a negative pressure. What is going to push the oil into the pump? The answer is *atmospheric pressure*. But how can this be?

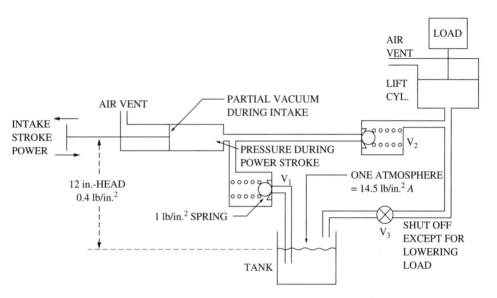

FIGURE 1.7
Basic pump with negative inlet pressure.

1.6   ATMOSPHERIC PRESSURE AND VACUUM

Although we live on the surface of the earth, we also live at the bottom of an ocean of air. The total weight of all this air is approximately 5.9 billion megatons, which is distributed relatively evenly over the surface of the earth. The weight of air directly above an area of 1 in.^{2} of the earth's surface, at sea level, on a clear day is approximately 14.7 lb, creating an absolute pressure of 14.7 lb/in.^{2}.

Note: Up to now we have studied pressure without considering the effects of the atmosphere. In other words, the atmosphere has been considered the zero reference point. The term (lb/in.^{2} gauge) or psig is used to designate that "atmosphere," and not vacuum, is considered the zero reference point. The term lb/in.^{2} (absolute), lb/in.^{2} (a), or psia is used when an "absolute vacuum" is considered as zero reference.

Figure 1.8 illustrates a uniform column of air with a 1-in.^{2} surface area extending all the way from the earth into space. The weight of this column is 14.7 lb. However, it would take only 30 in. of mercury (29.92 in., to be exact) to weigh the same amount as this full column of air.

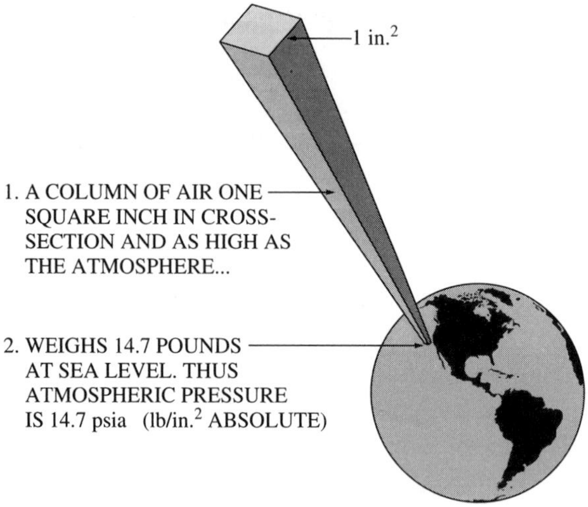

FIGURE 1.8
Atmospheric pressure.

1. A COLUMN OF AIR ONE
SQUARE INCH IN CROSS-
SECTION AND AS HIGH AS
THE ATMOSPHERE...

2. WEIGHS 14.7 POUNDS
AT SEA LEVEL. THUS
ATMOSPHERIC PRESSURE
IS 14.7 psia (lb/in.^{2} ABSOLUTE)

—1 in.^{2}

Mercury Barometer

Figure 1.9 shows a mercury barometer used to measure atmospheric pressure. To set up
the barometer, a glass tube (more than 30 in. tall) is completely filled with mercury, then
turned upside down into a small reservoir of mercury without allowing any air into the
tube.

FIGURE 1.9
Mercury barometer.

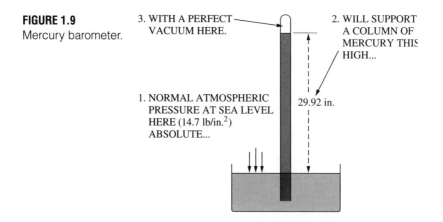

3. WITH A PERFECT
VACUUM HERE.

2. WILL SUPPORT
A COLUMN OF
MERCURY THIS
HIGH...

1. NORMAL ATMOSPHERIC
PRESSURE AT SEA LEVEL
HERE (14.7 lb/in.^{2})
ABSOLUTE...

29.92 in.

Some of the mercury will move down out of the tube into the reservoir, leaving
exactly 29.92 in. of mercury in the tube above the surface of the reservoir (assuming that
the barometer is at sea level and the atmosphere is normal) and a *perfect vacuum* in the
upper part of the tube above the mercury. The column of mercury is kept up in the tube
by the atmospheric pressure, normally 14.7 lb/in.^{2}, pushing down on the surface of the
mercury in the reservoir. The column of mercury in the tube will rise or fall, registering
any change in the barometric pressure of the atmosphere by showing more or less mercury height needed to counterbalance the atmospheric pressure.

Atmospheric pressure is measured relative to a perfect vacuum. The units are either $lb/in.^2$ *absolute (sometimes called psia) or "inches of mercury."*

Refer again to the pumping circuit of Figure 1.7. This circuit is the same as that of Figure 1.6 except that the tank is mounted 1 ft below the pump inlet. In order for the fluid to get into the pump, it must be lifted 1 ft and pushed through the check valve V_1. This will require 1.4 $lb/in.^2$ (0.4 $lb/in.^2$ to lift the fluid 1 ft and 1 $lb/in.^2$ to push the oil through V_1).

Creating a Vacuum in the Pump

The tank of Figure 1.7, like the mercury reservoir of the barometer in Figure 1.9, has the atmospheric pressure of about 14.5 $lbs/in.^2$ (a) pushing down on the surface of the oil. All we have to do is create a sufficient *partial vacuum* in the pumping chamber and the atmospheric pressure will push the fluid up the 1-ft distance, through the check valve, and into the pump.

Note: The atmospheric pressure at sea level with ideal weather conditions is 14.7 $lb/in.^2$ (a) (a = absolute). However 14.5 $lb/in.^2$ is assumed as nominal and is used as a unit of pressure in the metric system. One bar = one atmosphere = 14.5 $lb/in.^2$ (a). See Chapter 2, Table 2.2.

We create this vacuum by pulling on the pump handle and expanding the volume capacity of the pumping chamber. Any pressure generated inside the pump that is lower than the atmospheric pressure on the tank fluid, 14.5 $lb/in.^2$ (a), is considered a partial vacuum.

Note: When the volume of a chamber in which a fluid is confined is increased, the pressure of the fluid will be reduced; and to lower pressure below the atmospheric level is to produce a vacuum. Many pumps will perform satisfactorily with this partial vacuum. However, excessive vacuum can result in a damaging condition called cavitation. *To prevent cavitation, the pump inlet should be limited to a vacuum of about −5 in. of mercury (about −2.5 $lb/in.^2$) or an absolute pressure of not less than 12 $lb/in.^2$ (a).*

1.7   SUMMARY

1. Head pressure is caused by the density and height of a fluid above the reference point.
2. Head pressure is proportional to the height of the fluid regardless of the shape of the container.
3. Pascal's law states (in effect): Pressure applied to a confined fluid is transmitted in all directions, and acts with equal force on equal areas, and at right angles to them.
4. Energy can neither be created nor destroyed: "Input energy equals output energy."
5. The hydraulic pump does not suck the oil into its inlet. The oil must be pushed into the inlet. Then, the pump can push it out.
6. A check valve allows flow in one direction, but blocks flow in the opposite direction.
7. Generally pressure is measured with a gauge and the atmospheric pressure is considered as reference zero. This measurement is sometimes labeled "$lb/in.^2$ gauge" or psig.

8. Atmospheric pressure is measured relative to a perfect vacuum. The units are either lb/in.^{2} absolute (sometimes called psia) or "inches of mercury."

9. Most pumps must create a partial vacuum in their inlets in order for oil to be pushed into it by the atmospheric pressure.

10. A partial vacuum is produced in the inlet of a pump when the volume capacity of the pumping chamber is expanded by the pumping action.

1.8   PROBLEMS AND QUESTIONS

1. List two units of pressure.
2. True or false?: Head pressure is halved when the height of the fluid is halved.
3. True or false?: Head pressure increases with volume if the height of the fluid is held constant.
4. True or false?: The term psig means absolute pressure.
5. List three cases where head pressures are significant.
6. What would have been the upward force from the underside of the basement floor of Figure 1.2 had the water risen to 4 ft in the standpipe?
7. What would have been the total pressure, including the head pressure, at a point 2 in. from the bottom of the jug of Figure 1.3b if the cork were pushed down with a force of 10 lb?
8. In the hydraulic leverage system of Figure 1.4b, how much weight could be supported by the large cylinder with 15 lb applied to the small one?
9. What law allows a small force, moving a long distance, to move a large load a short distance?
10. How much movement of the pump cylinder in Figure 1.6 is required to move the lift cylinder 1 in.?
11. How many inches of mercury (in. Hg) are there in one lb/in.^{2}?
12. What pressure would be at the bottom of a 6-ft column of water, open to the atmosphere at the top?
13. Vacuum is measured in what units?
14. If the stand tube of the barometer in Figure 1.9 were heightened and the mercury replaced with water, how far above the reservoir would the water rise in the tube? (*Hint:* Assume the head pressure of water to be 0.4 lb/in.^{2} per foot.)
15. How much work is required at the pump of Figure 1.6 to lift the load 10 in.?

2
Dimensional Calculations

When analyzing hydraulics or other mechanical systems involving dimensions such as length, area, volume, and time, it is good practice to apply dimensional analysis to their solutions. This technique assures the proper use of conversion factors and virtually eliminates the need to memorize formulas. With all dimensions properly arranged in the statement of a problem, the solution consists merely of performing the indicated operations of the numbers and combining exponents of like bases. This method allows cancelation of all dimensions not needed, while keeping those called for in the answer.

   Although formulas are not needed in the solution of dimensional problems, they are generally shown in this text in order to ease your transition from formula to dimensional calculations. But, by all means, do not continue to use a memorized formula as a crutch. It is hoped that before you finish this book you can leave your "little black book" of formulas at home.

EXAMPLE:

How far will a 3-in.^{2} piston move if it receives 30 in.^{3} of fluid?

Solution: Let's examine the problem. The answer we need is the distance traveled (in.). We are given the volume of oil supplied to the cylinder (in.^{3}) and the area of the cylinder (in.^{2}). By dividing in.^{3} by in.^{2} we get in.^{3}/in.^{2} = in.$^{3-2}$ = in.^{1} = in.

   The formula for distance traveled is:

$$\text{Distance} = \text{Volume} \times \frac{1}{\text{Piston area}}$$

$$= 30 \text{ in.}^3 \times \frac{1}{3 \text{ in.}^2}$$

$$= \frac{30}{3} \text{ in.}^{3-2} = 10 \text{ in.}^1 = \textbf{10 in.}$$

Note: The formula for distance could have been stated as:

$$\text{Distance} = \frac{\text{Volume}}{\text{Area}}$$

While dividing volume by area will in fact give us distance, it is better generally to keep everything in the multiplying mode. That is, multiply by $\frac{1}{area}$ instead of dividing by area.

First, combine exponents of like bases in.^{3}/in.^{2} = in.$^{3-2}$ = in.^{1} = **in.** Since the answer has the proper dimension you will know that you have the dimensions properly placed. Next, perform the indicated operations of the numbers (30/3 = 10). The total answer is the combination of the two answers, or **10 in.**

A general shortcut used for combining like bases is to apply strikethroughs to show divisions or cancelations of exponents from bottom to top or top to bottom of the fraction. For example,

$$\text{Distance} = \underset{\text{10 in.}}{\cancel{30 \text{ in.}^3}} \times \frac{1}{\cancel{3 \text{ in.}^2}}$$

$$= \textbf{10 in.}$$

This is read "Inches cubed divided by inches squared equals inches; and 30 divided by 3 equals ten."

To use the distance formula, without the dimensions, is to apply the "rote method," which generally leads to trouble and errors. *There is no way that in.^{3} and in.^{2} can be combined to provide in. other than to divide in.^{3} by in.^{2}.* If the dimensions had been arranged differently in the solution, say, in.^{2}/in.^{3}, the answer would have been in.$^{2-3}$ = in.$^{-1}$. Then we would have realized immediately that the dimensions have to be rearranged in order to obtain the proper answer.

With the use of dimensional calculations you are constantly checking your answer as you cancel unwanted dimensions. You are learning from the procedure instead of blindly applying a memorized formula.

2.1 REVIEW OF DIMENSIONS

Figure 2.1 provides a review of some common dimensions.

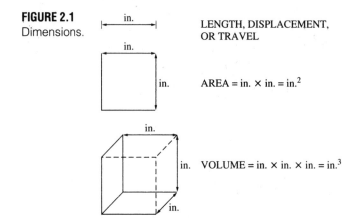

FIGURE 2.1
Dimensions.

LENGTH, DISPLACEMENT, OR TRAVEL

AREA = in. × in. = in.^{2}

VOLUME = in. × in. × in. = in.^{3}

TABLE 2.1

Common conversion fractions using U.S. units (see Section 2.2: Conversion Fractions)

U.S. Units	Conversion Fractions	Descriptions
Displacement (revolutions and radians)		
rev = 6.28 rad		
	$\dfrac{\text{rev}}{6.28\text{ rad}} = 1$	rev = revolution
	$\dfrac{6.28\text{ rad}}{\text{rev}} = 1$	rad = radian
	$\dfrac{\text{rev}}{\text{min}} \times \dfrac{6.28\text{ rad}}{\text{rev}} \times \dfrac{\text{min}}{60\text{ sec}} = 0.1046\text{ rad/sec}$	
	$\dfrac{\text{rev/min}}{0.1046\text{ rad/sec}} = 1$	sec = seconds
	$\dfrac{0.1046\text{ rad/sec}}{\text{rev/min}} = 1$	min = minutes
Flow		
1 gal/min = 231 in.^{3}/min		in.^{3}/min = cubic inches per minute
	$\dfrac{231\text{ in.}^3\text{/min}}{1\text{ gal/min}} = 1$	gal/min = gallons per minute
	$\dfrac{1\text{ gal/min}}{231\text{ in.}^3\text{/min}} = 1$	
Length (feet and inches)		
12 in. = 1 ft		
	12 in./1 ft = 1	in. = inches
	1 ft/12 in. = 1	ft = feet
Power (horsepower and kilowatts)		
hp = 0.746 kW		
	hp/0.746 kW = 1	kW = kilowatts
	0.746 kW/hp = 1	hp = horsepower
Pressure [lb/in.^{2} and in. of mercury (in. Hg)]		
29.92 in. Hg = 14.7 lb/in.^{2}		
	$\dfrac{14.7\text{ lb/in.}^2}{29.92\text{ in. Hg}} = 1$	lb = pounds in. = inches
	$\dfrac{29.92\text{ in. Hg}}{14.7\text{ lb/in.}^2} = 1$	Hg = mercury
Time (hours, minutes, and seconds)		
60 min = 1 h		
	60 min/1 h = 1	h = hours
	1 h/60 min = 1	
60 sec = 1 min		
	60 sec/1 min = 1	sec = seconds
	1 min/60 sec = 1	min = minute
Volume (in.^{3} and gallons)		
231 in.^{3} = 1 gal		
	231 in.^{3}/1 gal = 1	in.^{3} = cubic inches
	1 gal/231 in.^{3} = 1	gal = gallons

2.2 CONVERSION FRACTIONS

When the value of the numerator (top) of a fraction is equal to the value of its denominator (bottom), the absolute value of the fraction is one, and because multiplying by 1 does not change the absolute value of an expression, dimensions can be converted at will with the use of the conversion fractions given in Tables 2.1, 2.2, and 2.3. Tables 2.4 and 2.5, respectively, show prefixes for forming SI units and some handy formulas.

TABLE 2.2
Common conversion fractions using SI units

SI Units	Conversion Fractions	Descriptions
Pressure (Pascals, kilopascals, and bars*)		
$Pa = N/m^2$		
	$Pa/(N/m^2) = 1$	Pa = Pascal
	$(N/m^2)/Pa = 1$	N = Newton (force)
1000 Pa = 1 kPa		m^2 = square meters
	1000 Pa/kPa = 1	
	kPa/1000 Pa = 1	kPa = kilopascal
1 bar = 100 kPa		
	bar/100 kPa = 1	bar = approximately 1
	100 kPa/bar = 1	atmosphere
Area (m^2, dm^2, cm^2, mm^2)		
One square meter equals 100 square decimeters:		
	$m^2/100\ dm^2 = 1$	dm^2 = square decimeter
	$100\ dm^2/m^2 = 1$	
One square meter equals 10,000 square centimeters:		
	$m^2/10,000\ cm^2 = 1$	cm^2 = square centimeter
	$10,000\ cm^2/m^2 = 1$	
One square meter equals 1,000,000 square millimeters:		
	$m^2/1,000,000\ mm^2 = 1$	mm^2 = square millimeter
	$1,000,000\ mm^2/m^2 = 1$	
Volume (Liter, kiloliter, m^3, and cm^3)		
$m^3 = 1000\ L$		m^3 = cubic meter
	$m^3/1000\ L = 1$	L = liter
	$1000\ L/m^3 = 1$	1000 L = kiloliter
liter = 1000 cm^3		
	$L/1000\ cm^3 = 1$	cm^3 = cubic centimeter
	$1000\ cm^3/L = 1$	

*One bar is approximately the average atmospheric pressure at the earth's surface (or 14.5 $lb/in.^2$).

TABLE 2.3

Common conversion fractions using both U.S. and SI units (converting from one system to the other)

Units	Conversion Fractions	Descriptions
Flow (gal/min and L/min)		
1 gal/min = 3.7854 L/min	$\dfrac{\text{gal/min}}{3.7854\ \text{L/min}} = 1$	gal/min = gallons per minute
	$\dfrac{3.7854\ \text{L/min}}{\text{gal/min}} = 1$	L/min = liters per minute
Force or Weight (pounds, newtons, and kilograms)		
1 lb* = 4.44822 N		
	lb/4.44822 N = 1	lb = pound
	4.44822 N/lb = 1	
1 kilogram (weight or force) = 9.806654453 newtons (weight or force)		
kg_f = 9.8066 N		
	kg_f/9.8066 N = 1	kg = kilogram mass or kilo-
	9.8066 N/kg_f = 1	gram weight
1 kilogram (force)† = 2.2 lb		
	kg_f/2.2 lb = 1	
	2.2 lb/kg_f = 1	
Length (meters and inches)	39.37 in./m = 1	in. = inches
1 m = 39.37 in.	1 m/39.37 in. = 1	m = meter
Pressure (kPa, and bar versus lb/in.^{2})		
6.895 kPa = 1 lb/in.^{2}		
	$\dfrac{6.895\ \text{kPa}}{\text{lb/in.}^2} = 1$	kPa = kilopascals
	$\dfrac{\text{lb/in.}^2}{6.895\ \text{kPa}} = 1$	
1 bar = 14.5 lb/in.^{2}		
	$\dfrac{\text{bar}}{14.5\ \text{lb/in.}^2} = 1$	lb/in.^{2} = pounds per square inch
	$\dfrac{14.5\ \text{lb/in.}^2}{\text{bar}} = 1$	
Square Measures (m^2 and in.^{2})		
1 m^2 = 1550 in.^{2}		
	$\dfrac{1550\ \text{in.}^2}{\text{m}^2} = 1$	in.^{2} = square inch
	$\dfrac{1\ \text{m}^2}{1550\ \text{in.}^2} = 1$	m^2 = square meter
Torque (lb-in.) = 0.112979 N-m		
Volume (L and gal)	gal/3.7854 L = 1	gal = gallons (U.S.)
3.7854 L = gal	3.7854 L/gal = 1	L = liter
Volume (m^3 and in.^{3})	61,024 in.^{3}/m^3 = 1	in.^{3} = cubic inches
1 m^3 = 61,024 in.^{3}	1 m^3/61,024 in.^{3} = 1	m^3 = cubic meter
Weight (see Force above)		

*Some experts use lb_f here to be consistent with kg_f versus newton. This is not necessary, however, because "lb" always means force or weight (never mass).

†The term kg_f is used here to indicate that we are using kilograms of weight or force and not of mass.

TABLE 2.4
Factors and prefixes for forming decimal multiples and submultiples of SI units

Factor by Which Unit Is Multiplied	Prefix	Symbol	Factor by Which Unit Is Multiplied	Prefix	Symbol
10^{12}	tera	T	10^{-2}	centi	c
10^{9}	giga	G	10^{-3}	milli	m
10^{6}	mega	M	10^{-6}	micro	μ
10^{3}	kilo	k	10^{-9}	nano	n
10^{2}	hecto	h	10^{-12}	pico	p
10	deca	da	10^{-15}	femto	f
10^{-1}	deci	d	10^{-18}	atto	a

TABLE 2.5
Handy formulas

Boyle's Law

$P_1V_1 = P_2V_2$

$V_2 = (P_1V_1)/P_2$

$P_2 = (P_1V_1)/V_2$

Where P_1 (psia) and V_1 are the original pressure and volume of gas; P_2 (psia) and V_2 are the gas pressure and volume after the pressure or volume has been changed.

% Efficiency

$$(\text{General}) = \frac{\text{Output}}{\text{Input}} \times 100$$

$$\text{Pump mechanical efficiency} = \frac{\text{Theoretical input torque}}{\text{Actual torque to drive}} \times 100$$

$$\text{Pump volumetric efficiency} = \frac{\text{Actual flow output}}{\text{Theoretical flow output}} \times 100$$

Pump overall efficiency = Mechanical efficiency × Volumetric efficiency

or

$$\text{Pump overall efficiency} = \frac{\text{Output horsepower}}{\text{Input horsepower}} \times 100$$

Perfect vacuum = 29.95 in. Hg (barometer corrected)

2.3   REVIEW OF EXPONENTS

When multiplying common bases containing exponents, merely add the exponents and keep the common base.

■ | **EXAMPLE:**

Using powers of ten:

$$100 \times 1000 = 10^2 \times 10^3 = 10^{2+3} = 10^5 = \mathbf{100{,}000}$$ ■

When dividing common bases containing exponents, merely subtract the exponent of the divisor from the exponent of the number being divided.

■ **EXAMPLE:**

$$\frac{100,000}{1000} = \frac{10^5}{10^3} = 10^{5-3} = 10^2 = \mathbf{100}$$

or (using strikethroughs)

$$\frac{10^{\cancel{5}}{}^{2}}{10^{\cancel{3}}} = 10^2 = \mathbf{100}$$

OTHER EXAMPLES:

$$\text{in.}^1 \times \text{in.}^2 = \text{in.}^{1+2} = \text{in.}^3 \text{ or}$$
$$\text{in.}^3/\text{in.}^1 = \text{in.}^{3-1} = \text{in.}^2$$

or (using strikethroughs)

$$\frac{\cancel{\text{in.}^{3}}{}^{2}}{\cancel{\text{in.}}} = \mathbf{in.^2}$$

■

Note: See Figure 2.1.

2.4   EXAMPLE PROBLEMS

■ **EXAMPLE 2.1:**

How much pressure is required of the pump in Figure 2.2 to support a 1000-lb weight if the active area of the cylinder is 2 in.^{2}?

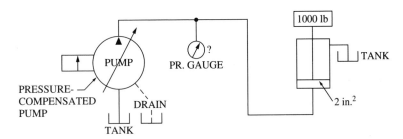

FIGURE 2.2
Hydraulic lift (U.S. units).

Solution: The formula for pressure is:

$$\text{Pressure} = \text{Force} \times (1/\text{Area})$$

We know that the dimensional units of the answer should be pounds per square inch or lb/in.^{2}, so we arrange the dimensions in this order (with lb on top and in.^{2} on the bottom) and solve their numerical values. That is,

$$1000 \ \text{lb}/2 \ \text{in.}^2 = \textbf{(500 lb/in.}^2\textbf{)}$$ ∎

*Note: The formula given on this solution is not needed to solve the problem, but is provided here as an aid for those who are using dimensions for the first time. First establish the order of dimensions (lb/in.^{2}), then combine the numbers (divide 1000 by 2 = **500**), and keep the dimensions (lb/in.^{2}).*

Note: When solving dimensionally, leave the units in their dimensional form. Do not use names or abbreviations such as pascals or psi.

∎ ## EXAMPLE 2.2

How fast will the cylinder of Figure 2.3 move the load if a flow of 200 in.^{3}/min (200 cubic inches per minute) is directed to the cylinder area of 2 in.^{2}?

Solution: The formula for cylinder speed is:

$$\text{Speed} = \text{Flow} \times \frac{1}{\text{Cylinder area}}$$

Using strikethroughs:

$$\text{Speed} = \frac{\overset{100 \ \text{in.}}{\cancel{200 \ \text{in.}^3}}}{\text{min}} \times \frac{1}{\cancel{2 \ \text{in.}^2}}$$

$$= \textbf{100 in./min} \ (100 \ \text{inches per minute})$$ ∎

Note: As before, the formula is not needed but is used here to provide for a smoother transition to dimensional calculations.

FIGURE 2.3
Cylinder speed versus flow
(in.^{3}/min).

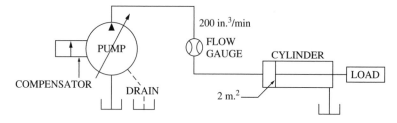

In order to obtain "in." in the answer we must put the in.^{3} on top and in.^{2} on the bottom of the fraction (in.^{3}/in.^{2}) = in.$^{3-2}$ = in.^{1} = in. To get in.^{2} on the bottom, we multiplied by 1/2 in.^{2}.

■ **EXAMPLE 2.3**

What is the speed of the cylinder of Figure 2.4 if it has an area of 30 in.^{2} and is receiving five gallons of fluid every minute (5 gal/min)?

Solution: The formula is the same: Speed = Flow × (1/Area), but now we have gallons to contend with.

$$\text{Speed} = \frac{5 \, \cancel{\text{gal}}}{\text{min}} \times \frac{231 \, \cancel{\text{in.}}^{\cancel{3}} \, \text{in.}}{\cancel{\text{gal}}} \times \frac{1}{30 \, \cancel{\text{in.}}^{\cancel{2}}}$$

$$= \textbf{38.5 in./min}$$

■

Note: Since 231 in.^{3} = 1 gal, multiplying by (231 in.^{3})/gal merely changes the dimension of flow without changing its absolute value.

FIGURE 2.4
Cylinder speed versus flow (gal/min).

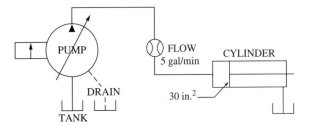

Suppose Example 2.3 were turned around to read: "How long will it take the load to travel 38.5 in.?"

Solution 1 Since we have already calculated the speed, the solution can be shortened. The formula would be:

$$\text{Time} = \text{Distance} \times \frac{1}{\text{Speed}}$$

$$= 38.5 \, \cancel{\text{in.}} \times \frac{1}{38.5 \, \cancel{\text{in.}}/\text{min}} = \text{min}$$

The numbers and inches cancel out, and min comes from the bottom of the bottom to the top, so:

$$\text{Time} = \textbf{1 min}$$

A better way to solve this problem and avoid the "bottom of the bottom" procedure is to start with the dimension on top that we want on top. For example, observe

$$\text{Time} = (\text{min/in.}) \times \text{in.}$$

$$\frac{1 \, \text{min}}{38.5 \, \cancel{\text{in.}}} \times 38.5 \, \cancel{\text{in.}}$$

$$= \textbf{1 min}$$

Note: Let's examine a "speed versus time" problem with which we are more familiar: Assume that we are driving our car at a fixed speed of 60 miles per hour, and we want to know how long it will take us to go 60 miles. The formula would be

$$\text{Time} = \text{Distance} \times \frac{1}{\text{Speed}} \quad \text{or}$$

$$= 60 \text{ miles} \times \frac{1}{60 \text{ miles/hour}}$$

Of course, the answer is **1 hour.** *The 60 miles on top cancels the 60 miles on the bottom, but the hours must come from the bottom of the bottom to the top to become part of the answer.*

A better way of solving the problem would have been to start with hours on top where it belongs or

$$\text{Time} = \frac{1 \text{ hour}}{60 \text{ miles}} \times 60 \text{ miles}$$

$$= \textbf{1 hour}$$

When solving dimensional problems, it doesn't matter whether we say 60 miles per hour or 1 hour per 60 miles. *Put the one you want on top. As long as you can cancel all other dimensions, the solution will be correct.*

Solution 2: Suppose we don't know the speed of the cylinder but we do know the information given in Example 2.3. The formula for time is:

$$\text{Time} = \text{Distance} \times \text{Area} \times (1/\text{Flow})$$

$$= 38.5 \text{ in.} \times 30 \text{ in.}^2 \times \frac{1}{5 \text{ gal/min} \times (231 \text{ in.}^3/\text{gal})}$$

$$= \textbf{1 min}$$

All dimensions except minutes are canceled, and min is on the bottom of the bottom, which brings it back to the top. The numbers combine to equal 1. So, the answer is:

$$\text{Time} = \textbf{1 min} \qquad ■$$

The best way to solve the preceding problem dimensionally is to start with the dimension on top that we want on top. The flow (5 gal/min), therefore, is turned upside down and becomes (min/5 gal). Likewise, (231 in.^{3})/gal is turned upside down and becomes gal/231 in.^{3}. We now have:

$$\text{Time} = \frac{\text{min}}{5 \text{ gal}} \times \frac{\text{gal}}{231 \text{ in.}^3} \times 38.5 \text{ in.} \times 30 \text{ in.}^2 = \textbf{1 min}$$

Note: Expressions like $\dfrac{\text{min}}{5 \text{ gal}}$ *(minute per 5 gallons) and* $\dfrac{5 \text{ gal}}{\text{min}}$ *(5 gallons per minute) both say what is happening in the pump: It takes 1 minute for the pump to deliver 5 gallons, or the pump delivers 5 gallons per minute.*

Any dimensional fraction may be inverted without changing its absolute value, provided that both the dimension and number are inverted together. (Observe that the 5 went with the gal when 5 gal/min was changed to min/5 gal).

■ **EXAMPLE 2.4**

Figure 2.5 shows a hydraulic system, using American National Standards Institute (ANSI) symbols, consisting of a pressure-compensated pump (P), a directional valve (V), and a cylinder (C). Certain parameters are to be determined from the information listed on the drawing.

    A. Determine the extending speed of cylinder C.

    B. What is the retracting speed of the cylinder?

    C. What is the flow from the rod end of the cylinder while extending?

Solution: Valve (V) is a three-position four-way valve, which (1) at the "center" position blocks the flow *from* the pump and *to* the tank; (2) at "extend," it directs flow *from* the pump *to* the piston end of the cylinder and *from* the rod end of the cylinder back *to* the tank (follow the arrows in the "extend" window of the valve symbol); and (3) at "retract," it directs flow *from* the pump *to* the rod end of the cylinder and from the piston end of the cylinder back *to* the tank (follow the arrows in the "retract" window of the valve symbol).

Note: More on directional valves is given in Chapter 4.

FIGURE 2.5
Directional control of cylinder
(U.S. units).

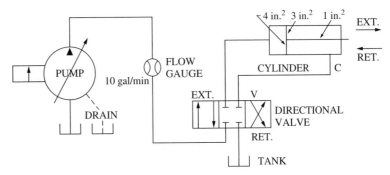

When valve (V) is at center, the pump (P) produces pressure at its compensated setting. At the other positions of the valve, the pump produces whatever pressure is required to move the load (up to a maximum of the compensator pressure setting) until the cylinder (C) "bottoms-out" in either direction; at which time, the pump returns to its pressure-compensated setting.

Note: The pressure-compensated pump is one method of limiting the pressure of a pumping system. Another system might use a relief valve for this purpose.

Solution A: During cylinder extension, we know:

1. The pump flow is 10 gal/min.
2. The area of the cylinder piston is 4 in.^{2}.
3. The formula for cylinder speed is

$$\text{Speed} = \text{Flow} \times \frac{1}{\text{Area}}$$

The *extending speed* of the cylinder is:

$$\text{Speed} = \frac{10 \text{ gal}}{\text{min}} \times \frac{231 \text{ in.}^3}{\text{gal}} \times \frac{1}{4 \text{ in.}^2}$$

$$= \textbf{577.5 in./min}$$

Note: To obtain in./min in the answer we must divide in.^{3} by in.^{2} to obtain in., and convert gallon to in.^{3} by multiplying by 231 in.^{3}/gal.

Strikethroughs are again used to show how exponents with like bases are combined.

Solution B: During cylinder retraction, we know:

1. The pump flow is 10 gal/min.
2. The effective area of the rod end (annulus area) of the cylinder is 4 in.^{2} − 1 in.^{2} = 3 in.^{2}.

Therefore, the *retracting speed* of the cylinder is:

$$\text{Speed} = \text{Flow} \times \frac{1}{\text{Area}}$$

$$= \frac{10 \text{ gal}}{\text{min}} \times \frac{231 \text{ in.}^3}{\text{gal}} \times \frac{1}{3 \text{ in.}^2}$$

$$= \textbf{770 in./min}$$

Solution C: During cylinder extension, we know:

1. The annulus area of the cylinder is 3 in.^{2}.
2. The extending speed of the cylinder is 577.5 in./min.

Therefore, the flow from the rod end is:

$$\text{Flow} = \text{Area} \times \text{Speed}$$

$$= 3 \text{ in.}^2 \times \frac{577.5 \text{ in.}}{\text{min}} \times \frac{\text{gal}}{231 \text{ in.}^3}$$

$$= \textbf{7.5 gal/min} \qquad\blacksquare$$

Note: The ratio of the annulus flow rate to the piston flow rate is the same as the ratio of the annulus area to the piston area $\left(\dfrac{7.5}{10} = \dfrac{3}{4}\right)$. It is logical that less oil will be pushed out of the cylinder by the annulus area than it took to displace the full piston area.

2.5   **USE OF SI UNITS IN DIMENSIONAL CALCULATION**

SI (Système International) units may be used in dimensional calculations in exactly the same way as the U.S. units provided that they are kept in their individual dimensions instead of the name that may have been assigned to a dimensional group. For example, use newtons per square meter (N/m^2) for pressure instead of pascals.

Directional Control of Cylinder (Metric SI Units)

The circuit of Figure 2.6 is the same as that of Figure 2.5 except the dimensions are expressed as SI units.

FIGURE 2.6
Directional control of cylinder (SI units).

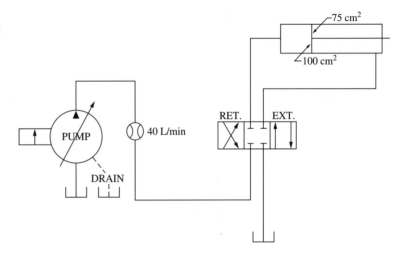

■   **EXAMPLE 2.5**

Determine the following from the information of Figure 2.6:

A. The extending speed of the cylinder (m/min)
B. The retracting speed of the cylinder (m/min)
C. The flow from the rod end of the cylinder while retracting (L/min).

Solution A: During cylinder extension we know:

1. The pump flow is 40 L/min.
2. The area of the cylinder piston is 100 cm^2.

Therefore, the *extending speed* of the cylinder is:

$$\text{Speed} = \text{Flow} \times \frac{1}{\text{Area}}$$

$$= \frac{40\cancel{L}}{\min} \times \frac{\overset{10}{\cancel{1000}}\,\text{cm}^{\cancel{3}}}{\cancel{L}} \times \frac{1}{\cancel{100}\,\text{cm}^{\cancel{2}}}$$

$$= \textbf{400 cm/min} \text{ or } \textbf{4 m/min}$$

Note: One liter = 1000 cm³ and 1000 cm³/L = 1.
Note: One meter equals 100 cm and 1 m/100 cm = 1. To change cm to m:

$$\frac{\overset{4}{\cancel{400 \text{ cm}}}}{\text{min}} \times \frac{1 \text{ m}}{\cancel{100 \text{ cm}}} = \textbf{4 m/min}$$

Solution B: During retraction we know:

1. The pump flow is 40 L/min.
2. The annulus area of the cylinder rod end is 75 cm².

Therefore, the *retracting speed* of the cylinder is:

$$\text{Speed} = \text{Flow} \times \frac{1}{\text{Area}}$$

$$= \frac{40 \cancel{L}}{\text{min}} \times \frac{1000 \text{ cm}^{\overset{\text{cm}}{3}}}{\cancel{L}} \times \frac{1}{75 \cancel{\text{cm}^2}}$$

$$= \textbf{533.3 cm/min}$$

or

$$= \textbf{5.333 m/min}$$

Solution C: During extension we know:

1. The annulus area of the piston is 75 cm².
2. The extending speed of the cylinder is 4 m/min.

Therefore, the *flow from the rod end while extending* is:

$$\text{Flow} = \text{Speed} \times \text{Area}$$

$$= \frac{400 \cancel{\text{ cm}}}{\text{min}} \times 75 \cancel{\text{ cm}^2} \times \frac{L}{1000 \cancel{\text{ cm}^3}}$$

$$= \textbf{30 L/min}$$ ∎

Note: One liter = 1000 cm³ and L/1000 cm³ = 1.

Hydraulic Lift (SI Units)

Figure 2.7 is similar to Figure 2.2 except that metric SI units are used.

FIGURE 2.7
Hydraulic lift (SI units).

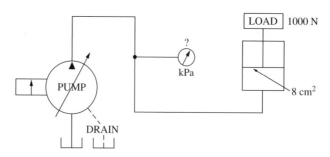

EXAMPLE 2.6

A pump is connected to an 8-cm^2 cylinder in order to lift a 100-N weight. What pressure is required of the pump?

Solution: Pressure $= \text{Force} \times \dfrac{1}{\text{Area}}$

$$= 1000 \text{ N} \times \frac{1}{(8 \text{ cm}^2) \times \text{m}^2/(10{,}000 \text{ cm}^2)}$$

$$= 1{,}250{,}000 \text{ N/m}^2$$

$$= 1{,}250{,}000 \text{ pascals}$$

$$= \textbf{1250 kPa (kilopascals)}$$

$$\text{or } 1250 \text{ kPa} \times \frac{(\text{lb/in.}^2)}{6.895 \text{ kPa}}$$

$$= \textbf{181.3 lb/in.}^2 \textbf{ (U.S. units)}$$

You can rest assured that, since your dimensions are right in the answer, the problem was set up properly. And if you did your arithmetic correctly you have the right answer. ■

Remember that you cannot use just any flow or speed when the areas of the cylinder are not equal. If the flow of concern is into or out of the cap end, you must use the full piston area. Likewise, if the flow of concern is into or out of the rod end, you must use the annulus area.

Note: The term cap end *may be used to identify the larger end of the cylinder with the full piston area.*

2.6   HORSEPOWER

EXAMPLE 2.7

Determine the hydraulic equivalent to mechanical horsepower. *Review:* Work is force acting through distance. The basic unit of work is the ft-lb, the work needed to lift one pound one foot. Power is rate of doing work, and can be expressed as ft-lb/min.

1 horsepower (hp) mechanical = 33,000 ft-lb/min

Note: Hydraulic power also shows the rate of doing hydraulic work. Hydraulic power is the product of pressure and flow in lb/in.^{2} × gal/min for U.S. units and kPa or bars × L/min for SI units. In fact the U.S. hydraulic horsepower equation is derived from the mechanical equation. Let's see how this is done.

Solution:

$$1 \text{ hp} = 33{,}000 \text{ ft-lb/min}$$

$$= \frac{33{,}000 \text{ ft-lb}}{\text{min}} \times \frac{12 \text{ in.}}{\text{ft}} \times \frac{\text{gal}}{\dfrac{231 \text{ in.}^3}{\text{in.}^2}}$$

$$= \frac{33,000}{231} \times 12 \times (\text{lb/in.}^2) \times (\text{gal/min}) \qquad \blacksquare$$

Another advantage of dimensional calculations is that any numerator (on top) can be relocated to any part of the expression as long as it remains on top. The same is true of the denominators (on the bottom). So, let's relocate some dimensions in order to obtain the hydraulic units lb/in.^{2} and gal/min in the new expression. We now have

$$1 \text{ hp} = (1714 \text{ lb/in.}^2) \times (\text{gal/min})$$

Note: In problems where fractions are being multiplied, any numerator (on top) may be exchanged for any other numerator. Likewise, the denominators (on the bottom) may be interchanged, all without changing the value of the answer. Examples: $\dfrac{3}{4} \times \dfrac{5}{8} = \dfrac{15}{32}$ *or* $\dfrac{3}{8} \times \dfrac{5}{4} = \dfrac{15}{32}$ *and* $\dfrac{1}{2} \times \dfrac{2}{3} = \dfrac{2}{6}$ *or* $\dfrac{2}{2} \times \dfrac{1}{3} = \dfrac{2}{6}$.

From this hydraulic power equation, we can formulate two conversion fractions:

$$\frac{1 \text{ hp}}{(1714 \text{ lb/in.}^2) \times (\text{gal/min})} = 1$$

$$\frac{(1714 \text{ lb/in.}^2) \times (\text{gal/min})}{1 \text{ hp}} = 1$$

EXAMPLE 2.8

Calculate the output horsepower of a pump if it delivers a flow of 10 gal/min and a pressure of 1000 lb/in.^{2}.

Solution

$$\text{Horsepower} = \underbrace{\frac{1 \text{ hp}}{(1714 \text{ lb/in.}^2) \times (\text{gal/min})}}_{\text{conversion fraction \#1}} \times 1000 \text{ lb/in.}^2 \times 10 \text{ gal/min}$$

$$= \textbf{5.83 hp} \qquad \blacksquare$$

Note: I wanted you to see how the dimensions of hydraulic horsepower were derived from mechanical horsepower. Now that you have understood these dimensional calculations for horsepower, and because the formula is so simple, consider the previous solution to be academic and feel free to use the following horsepower formula:

$$(\text{hydraulic}) \text{ hp} = \frac{(\text{lb/in.}^2) \times (\text{gal/min})}{1714} \quad \text{or} \quad \frac{\text{psi} \times \text{gal/min}}{1714}$$

$$= \frac{\text{Pressure (psi)} \times \text{Flow (gal/min)}}{1714}$$

$$= \frac{1000 \times 10}{1714}$$

$$= \textbf{5.83 hp}$$

2.7 U.S. AND SI UNITS FOR POWER

Power is expressed in kilowatts (kW = 1000 W) in SI units instead of horsepower. Regardless of the system used, remember that power equals pressure times flow times a factor used to obtain the desired power unit. Just as 1/1714 was used to obtain horsepower from the U.S. units (lb/in.^{2}) × gal/min, 1/600 must be used to change (bars × L/min) to kW; or

$$kW = \frac{bar \times L/min}{600}; \text{ and since 1 bar = 100 kPa}$$

$$\mathbf{kW = \frac{kPa \times L/min}{60,000}}$$

Tables 2.6 and 2.7 give power formulas for U.S. and SI units, respectively.

TABLE 2.6
U.S. unit power formulas

Units	Descriptions
Linear Mechanical Horsepower	**hp = Horsepower**
$\dfrac{ft\text{-}lb}{min}$ = 1 lb lifted 1 ft in 1 min	lb = pound (weight or force)
$hp = \dfrac{lb \times ft/min}{33,000}$	ft = foot = 12 in. min = minutes
Rotary Mechanical Horsepower	**(lb-in.) = torque**
$hp = \dfrac{(rev/min) \times (lb\text{-}in.)}{63,025}$	rev/min = revolutions per minute
Electrical Horsepower	**E = volts**
$hp = \dfrac{E \times I}{746}$	I = amps $E \times I$ = watts kW = kilowatts = 1000 W
Hydraulic Horsepower	
$hp = \dfrac{psi \times (gal/min)}{1714}$	psi = lb/in.^{2}

TABLE 2.7
SI unit power formulas

Units	Descriptions
Linear Mechanical Power	
$1\text{ W} = 1$ newton lifted 1 meter in 1 second or $1\text{ W} = \dfrac{1\text{ N-m}}{\text{sec}}$	
$W = \dfrac{\text{N-m/min}}{60}$	W = watt unit of SI power N = newton unit of SI force m = meter min = minute
$kW = \dfrac{\text{N-m/min}}{60{,}000}$	kW = kilowatt = 1000 W unit of SI power
Rotary Mechanical Power	
$kW = \dfrac{\text{N-m} \times (\text{rev/min})}{9550}$	Torque = N-m = newton-meter rev/min = revolutions per minute L = liter
Hydraulic Power	
$kW = \dfrac{(\text{L/min}) \times \text{bar}}{600}$	bar = atmosphere unit of SI pressure
$kW = \dfrac{(\text{L/min}) \times \text{kPa}}{60{,}000}$	kPa = kilopascal = 1000 pa, *another* unit of SI pressure

2.8   SUMMARY

1. The use of dimensional calculations virtually eliminates the need for formulas and ensures proper dimensions in the answer.
2. The conversion fraction has an absolute value of one.
3. When multiplying like bases, add exponents.
4. When dividing like bases, subtract exponents.
5. Leave the units in their dimensional form. Do not use abbreviations such as psi (except in the horsepower formula, where its use is acceptable).
6. Multiply by 1/2 instead of dividing by 2.
7. Expressions like min/5 gal and 5 gal/min both say what is happening in a pump: It takes 1 minute to deliver 5 gallons or it delivers 5 gallons per minute.
8. Any dimensional fraction may be inverted without changing its absolute value, provided that both the dimensions and numbers are inverted together.
9. In problems where fractions are being multiplied together, any numerator (on top) may be exchanged for any other numerator. Likewise, the denominators (on the bottom) may be interchanged, all without changing the value of the answer.
10. With a fixed flow from the pump, a single-rod cylinder will retract faster than it extends.

11. Another advantage of dimensional calculations is that any numerator (on top) can be relocated to any part of the expression as long as it remains on top. The same is true of the denominators (on the bottom).
12. There is no way that in.^{3} and in.^{2} can be combined to provide in. other than to divide in.^{3} by in.^{2}.
13. SI units may be used in dimensional calculations in exactly the same way as U.S. units provided that they are kept in their individual dimensions. No names or abbreviations may be used for dimensions. For example, use newtons per square meter (N/m^2) for pressure instead of pascals.

2.9   PROBLEMS AND QUESTIONS

1. How far will a 5-in.^{2} cylinder move if it receives 30 in.^{3} of fluid?
2. True or false?: When the value of the numerator (top) of a fraction is equal to the value of its denominator (bottom), the absolute value of the fraction is one.
3. True or false?: When dividing common bases containing exponents, merely add the exponent of the divisor to the exponent of the dividend (number being divided).
4. How much pressure is required of the pump in Figure 2.8 to support a 1500-lb weight, if the active area of the cylinder is 3 in.^{2}?

FIGURE 2.8
Hydraulic lift.

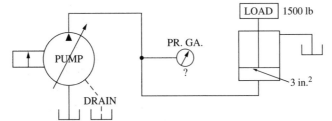

5. How fast will the cylinder of Figure 2.9 move the load if a flow of 300 in.^{3}/min (300 cubic inches per minute) is directed to the cylinder area of 5 in.^{2}?

FIGURE 2.9
Cylinder speed versus in.^{3}/min.

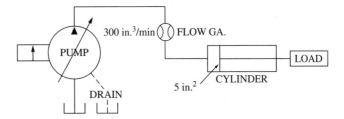

6. How fast will the cylinder of Figure 2.10 move the load if a flow of 10 gal/min is directed to the cylinder area of 12 in.²?

FIGURE 2.10
Cylinder speed versus gal/min.

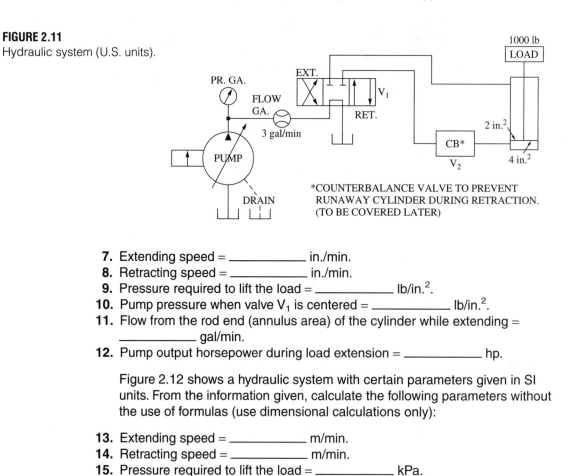

Figure 2.11 shows a hydraulic system with certain parameters given in U.S. units. From the information given, calculate the following parameters without the use of formulas (use dimensional calculations only):

FIGURE 2.11
Hydraulic system (U.S. units).

7. Extending speed = ＿＿＿＿＿ in./min.
8. Retracting speed = ＿＿＿＿＿ in./min.
9. Pressure required to lift the load = ＿＿＿＿＿ lb/in.².
10. Pump pressure when valve V_1 is centered = ＿＿＿＿＿ lb/in.².
11. Flow from the rod end (annulus area) of the cylinder while extending = ＿＿＿＿＿ gal/min.
12. Pump output horsepower during load extension = ＿＿＿＿＿ hp.

Figure 2.12 shows a hydraulic system with certain parameters given in SI units. From the information given, calculate the following parameters without the use of formulas (use dimensional calculations only):

13. Extending speed = ＿＿＿＿＿ m/min.
14. Retracting speed = ＿＿＿＿＿ m/min.
15. Pressure required to lift the load = ＿＿＿＿＿ kPa.

FIGURE 2.12
Hydraulic system (SI units).

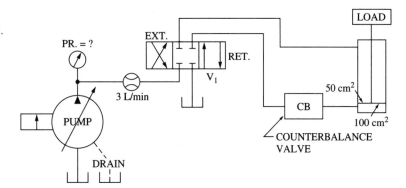

16. Pump pressure when valve V_1 is centered = _____ kPa.
17. Flow from the rod end (annulus area) of the cylinder while extending = _____ L/min.
18. Pump output power during load extension = _____ kW.
19. Calculate the (U.S.) horsepower output of a pump that supplies 5 gal/min flow at a pressure of 1500 lb/in.^{2}.
20. Convert the horsepower of Problem 19 to kW.
21. Calculate the SI output power of a pump in kilowatts, if the pressure is 100 bars and the flow is 12 liters per minute.
22. Calculate the SI output of a pump in kilowatts, if the flow is 10 liters per minute and the pressure is 10,000 kPa.

3
Hydraulic Actuators

This chapter considers the output members of the hydraulic system, the cylinder and motor. These units are also called actuators since they actuate or drive the load.

3.1   CYLINDERS

Cylinders are linear actuators, in that the output of a cylinder is straight-line motion. There are many classifications of cylinders.

Single-Acting Cylinders

Figure 3.1 shows ram-type or single-acting cylinders. They have only one port and exert force in only one direction. Most single-acting cylinders are mounted vertically, and are retracted by the weight of the load.

FIGURE 3.1
Ram-type cylinders are single acting.

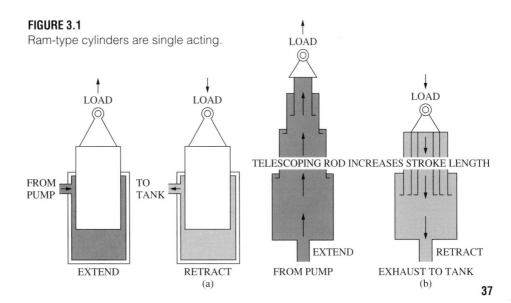

Ram-type cylinders are practical for long strokes and are used on jacks, automobile hoists, and elevators. The rod can be telescoped, as shown in Figure 3.1b, in order to extend travel while keeping a more compact retracted package.

Double-Acting Cylinders

Figure 3.2 shows a single-rod, double-acting cylinder with two power strokes. Flow may be directed to either end of the cylinder and exhausted from the other. The area of the rod reduces the effective area of the cylinder in one direction (retraction). This type is called a differential cylinder because of the difference in the working areas of the piston end and the rod end.

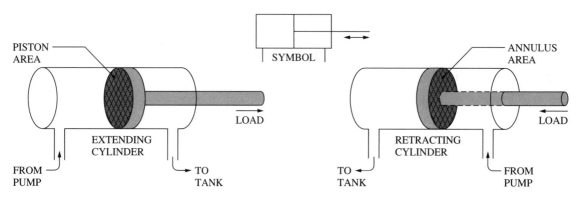

FIGURE 3.2
Single-rod double-acting cylinder.

■ **EXAMPLE 3.1**

(Use U.S. units; see Table 2.1 for conversion fractions.) What will be the speed of a cylinder when a fixed flow (say, 100 in.³/min) is directed to the rod end with an annulus area of, say, 2 in.²?

Solution: The formula for cylinder speed is:

$$\text{Speed} = \text{Flow} \times \frac{1}{\text{Area}}$$

$$= \frac{\overset{50 \text{ in.}}{\cancel{100 \text{ in.}^3}}}{\text{min}} \times \frac{1}{\cancel{2 \text{ in.}^2}}$$

$$= \textbf{50 in./min}$$

Note: The strikethroughs show that $\dfrac{\frac{in.^3}{in.}}{in.^2} = in.$, leaving the dimensions in./min and the

numbers $\dfrac{\frac{50}{100}}{\frac{2}{1}} = 50.$

If the same fixed flow of 100 in.³/min were directed to the other end of the cylinder, which has a full piston area of, say, 4 in.², the same cylinder would extend at a speed of:

$$Speed = Flow \times \frac{1}{Area}$$

$$= \frac{\overset{25\ in.}{\cancel{100}\ \cancel{in.^3}}}{min} \cdot \frac{1}{\cancel{4}\ \cancel{in.^2}}$$

$$= \textbf{25 in./min} \qquad ■$$

EXAMPLE 3.2

(Use SI units; see Table 2.2 for conversion fractions.) If a flow of 6 L/min were directed to the rod end of the cylinder with an annulus area of 5 cm², the cylinder will retract at a speed of:

$$Speed = Flow \times \frac{1}{Area}$$

$$= \frac{6\ \cancel{L}}{\cancel{min}} \times \frac{1000\ \overset{cm}{\cancel{cm^3}}}{\cancel{L}} \times \frac{1}{5\ \cancel{cm^2}}$$

$$= \textbf{1200 cm/min}$$

$$or \quad \frac{1200\ \cancel{cm}}{min} \times \frac{m}{100\ \cancel{cm}} = \textbf{12 m/min}$$

If the same flow of 6 L/min were directed to the cap end of the cylinder, which has an area of 10 cm², the cylinder will extend at a speed of:

$$Speed = Flow \times \frac{1}{Area}$$

$$= \frac{6\ \cancel{L}}{min} \times \frac{1000\ \overset{cm}{\cancel{cm^3}}}{\cancel{L}} \times \frac{1}{10\ \cancel{cm^2}}$$

$$= \textbf{600 cm/min}$$

So, with a fixed flow from the pump, the single-rod differential cylinder will retract faster than it extends. ■

Note: Chapter 6, Section 6.4, shows what happens to a differential cylinder when the pressure is fixed and the flow is allowed to swing inversely with restriction.

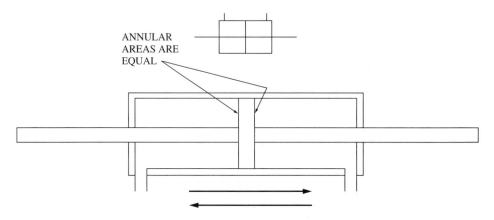

FIGURE 3.3
Double-rod cylinder with equal areas.

Figure 3.3 shows a double-acting, double-rod cylinder. This is not a differential cylinder when the two rods are the same size. The speed will be the same in either direction with a fixed flow rate.

Note: A double-acting cylinder may be used as a single-acting unit by draining the inactive end to tank.

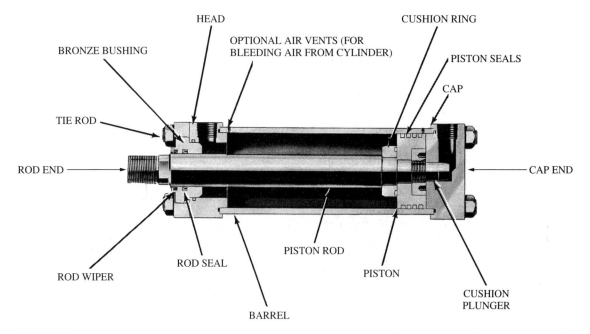

FIGURE 3.4
Cylinder cutaway. (Courtesy of Vickers Inc.)

Cylinder Construction

Figure 3.4 is a cutaway drawing showing the essential parts of a cylinder. The barrels are usually made of seamless steel tubing, honed to a fine finish on the inside. Pistons are usually made of steel or cast iron, with seals provided to reduce the internal leakage between piston and barrel. Rod seals, inserted into bronze bushings, prevent external leakage.

Cylinder Leakage

V-type and O-ring rod seals are generally used to prevent external leaks with reasonable friction losses, and they do a satisfactory job.

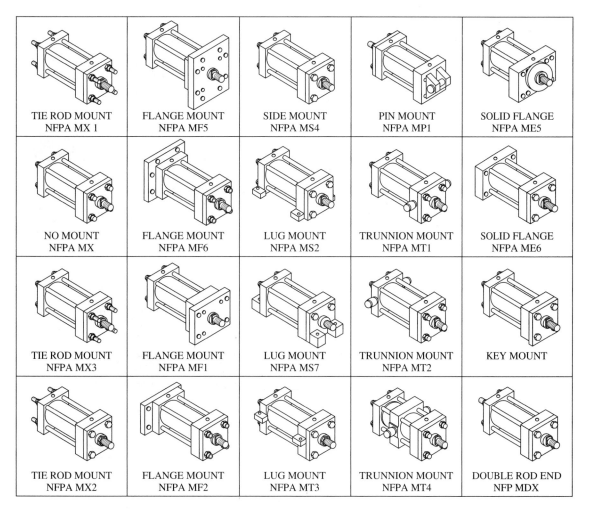

FIGURE 3.5
Cylinder mountings. (Courtesy of Milwaukee Cylinder Company, Milwaukee, Wisc.)

Where very low internal leakage is required, such as to hold a load in a nonservo system, the piston seal may consist of an O-ring or T-ring with two or more heavy-duty backup rings. Where precision is needed, and a small amount of leakage can be tolerated, such as in a servo-controlled mechanism, the piston seal may consist of an automotive-type piston ring. This type of seal usually provides better servo accuracy because it requires less break-away force.

Cylinder Mounting

Figure 3.5 shows many mounting configurations for cylinders. You may choose the right one for your particular application. For servo applications, tighten (torque) the tie-bolts to the maximum of their recommended stress limits. This increases the stiffness of the cylinder, providing better servo control.

Note: Servo control is covered in Chapter 9.

FIGURE 3.6
Cylinder cushion.

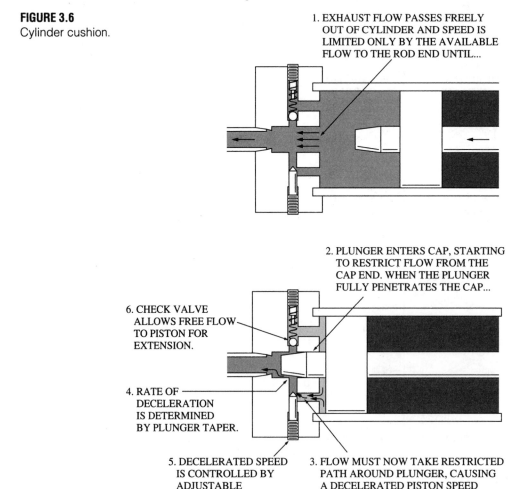

1. EXHAUST FLOW PASSES FREELY OUT OF CYLINDER AND SPEED IS LIMITED ONLY BY THE AVAILABLE FLOW TO THE ROD END UNTIL...

2. PLUNGER ENTERS CAP, STARTING TO RESTRICT FLOW FROM THE CAP END. WHEN THE PLUNGER FULLY PENETRATES THE CAP...

6. CHECK VALVE ALLOWS FREE FLOW TO PISTON FOR EXTENSION.

4. RATE OF DECELERATION IS DETERMINED BY PLUNGER TAPER.

5. DECELERATED SPEED IS CONTROLLED BY ADJUSTABLE OPENING.

3. FLOW MUST NOW TAKE RESTRICTED PATH AROUND PLUNGER, CAUSING A DECELERATED PISTON SPEED UNTIL THE PISTON BOTTOMS OUT.

Cylinder Cushions

Cushions are often added to either or both ends of the cylinder to decelerate the load to a slower speed near the end of travel. Figure 3.6 shows a cushion added to the cap end of a cylinder. Retraction speed is normal until the tapered plunger enters the cap and starts to restrict flow from the barrel to the port. The deceleration rate is determined by the plunger taper angle, and the reduced speed is determined by the adjustable restriction in the cylinder cap.

Cylinder Application Examples (U.S. Units)

Figure 3.7 shows application data, in U.S. units, for cylinders with $1\frac{1}{2}$- to 8-in. bore sizes. The following examples will help you become acquainted with the table.

■ | **EXAMPLE 3.3**

Suppose you want to lift a 3000-lb load with a pressure of 1000 lb/in.^{2}. Which cylinder would you select from the table of Figure 3.7?

Solution: The formula for cylinder area is:

$$\text{Area} = \text{Force (weight)} \times \frac{1}{\text{Pressure}}$$

where 1/Pressure = inverted pressure or in.^{2}/lb. Therefore,

$$\text{Area} = \cancel{lb} \times \frac{\text{in.}^2}{\cancel{lb}} = \textbf{in.}^2$$

The smallest possible cylinder area size would be:

$$\text{Area} = \overset{3}{\cancel{3000}}\,\cancel{lb} \times \frac{\text{in.}^2}{\underset{}{\cancel{1000}\,\cancel{lb}}} = \textbf{3 in.}^2$$

Note: $\frac{1}{Pressure}$ *expressed with units becomes* $\frac{1}{lb/in.^2}$. *We can bring the fraction lb/in.^{2} from the bottom of the equation to the top by inverting it. So, the term* $\frac{1}{lb/in.^2}$ *can be written as in.^{2}/lb.*

The term "lb" cancels "lb" as shown by the strikethroughs, leaving only the dimension "in.^{2}" and the numbers (3000/1000 = 3).

From the table of Figure 3.7, the smallest cylinder with a piston area larger than 3 in.^{2} is the one with a 2-in. bore, and an area of 3.142 in.^{2}. *Normally this is the one you would select.*   ■

■ | **EXAMPLE 3.4**

How fast would the 2-in. bore cylinder extend if it is supplied with a flow of 5 gal/min?

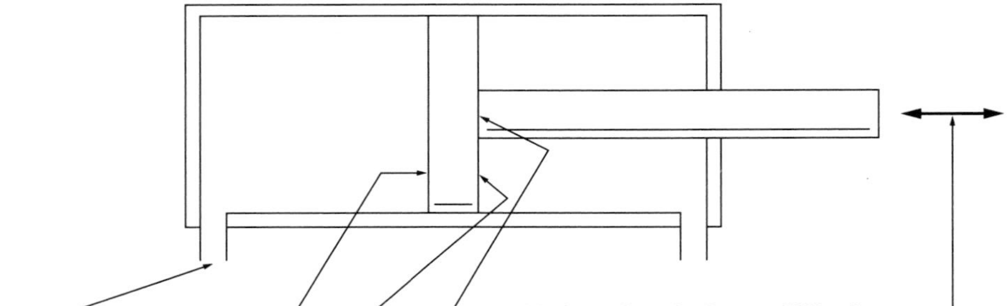

(Maximum Operating Pressure 2000 psi)

CYLINDER BORE	PORT SIZE: N.P.T. THREAD	PORT SIZE: *STRAIGHT THREAD	*ROD O.D.	**PISTON AREA (SQUARE INCH): FULL BORE	**PISTON AREA: ANNULUS	**PISTON AREA: ROD	RATIO FULL BORE TO ANNULUS AREA	†APPROXIMATE OUTPUT FORCE—POUNDS: 500 PSI PUSH	500 PSI PULL	1000 PSI PUSH	1000 PSI PULL	1500 PSI PUSH	1500 PSI PULL	2000 PSI PUSH	2000 PSI PULL
1 1/2	1/2″	5/8″ TUBE OD (7/8-14 THD.)	5/8″ STD.	1.767	1.460	.307	1.21/1.00	884	730	1767	1460	2651	2190	3534	2920
			1″ HVY.		.982	.785	1.80/1.00		491		982		1473		1964
2	1/2″	5/8″ TUBE OD (7/8-14 THD.)	1″ STD.	3.142	2.357	.785	1.33/1.00	1571	1178	3142	2357	4713	3535	6284	4714
			1-3/8″ HVY.		1.657	1.485	1.90/1.00		828		1657		2485		3314
2 1/2	1/2″	3/4″ TUBE OD (1-1/16-12 THD.)	1″ STD.	4.909	4.124	.785	1.19/1.00	2455	2062	4909	4124	7364	6186	9818	8248
			1-3/8″ INT'MED.		3.424	1.485	1.43/1.00		1712		3424		5136		6848
			1-3/4″ HVY.		2.504	2.405	1.96/1.00		1252		2504		3756		5008
3 1/4	3/4″	3/4″ TUBE OD (1-1/16-12 THD.)	1-3/8″ STD.	8.296	6.811	1.485	1.22/1.00	4148	3405	8296	6811	12444	10216	16592	13622
			1-3/4″ INT'MED.		5.891	2.405	1.41/1.00		2945		5891		8836		11782
			2″ HVY.		5.154	3.142	1.61/1.00		2577		5154		7731		10308
4	3/4″	3/4″ TUBE OD (1-1/16-12 THD.)	1-3/4″ STD.	12.566	10.161	2.405	1.24/1.00	6283	5080	12566	10161	18849	15241	25132	20322
			2″ INT'MED.		9.424	3.142	1.33/1.00		4712		9424		14136		18848
			2-1/2″ HVY.		7.666	4.900	1.64/1.00		3833		7666		11500		15332
5	3/4″	1″ TUBE OD (1-5/16-12 THD.)	2″ STD.	19.635	16.493	3.142	1.19/1.00	9818	8246	19635	16493	29453	24739	39270	32986
			2-1/2″ INT'MED.		14.735	4.900	1.33/1.00		7367		14735		22102		29470
			3-1/2″ HVY.		10.014	9.621	1.96/1.00		5007		10014		15021		20028
6	1″	1″ TUBE OD (1-5/16-12 THD.)	2-1/2″ STD.	28.274	23.374	4.900	1.21/1.00	14137	11687	28274	23374	42411	35061	56548	46748
			3-1/2″ INT'MED.		18.653	9.621	1.52/1.00		9326		18653		27979		37306
			4″ HVY.		15.708	12.566	1.80/1.00		7854		15708		23562		31416
7	1-1/4″	1-1/2″ TUBE OD (1-7/8-12 THD.)	3″ STD.	38.485	31.416	7.069	1.23/1.00	19242	15708	38485	31416	57728	47124	76970	62832
			4″ INT'MED.		25.919	12.566	1.48/1.00		12959		25919		38878		51838
			5″ HVY.		18.850	19.635	2.04/1.00		9425		18850		28275		37700
8	1-1/2″	1-1/2″ TUBE OD (1-7/8-12 THD.)	3-1/2″ STD.	50.265	40.644	9.621	1.24/1.00	25133	20332	50265	40644	75398	60966	100530	81288
			4-1/2″ INT'MED.		34.361	15.904	1.46/1.00		17180		34361		51541		68722
			5-1/2″ HVY.		26.507	23.758	1.90/1.00		13253		26507		39760		53014

*STRAIGHT THREAD CONNECTIONS AVAILABLE UPON REQUEST.

†PULL FORCE VALUES APPLY IN BOTH DIRECTIONS FOR CYLINDERS WITH DOUBLE-ENDED PISTON RODS.

NOTE ONE: FLUID DISPLACEMENT ($in.^3$) IS EQUAL TO ACTIVE AREA ($in.^2$) TIMES CYLINDER DISPLACEMENT (in.) OR $in. \times in.^2 = in.^3$.

NOTE TWO: AREA = $\pi D^2/4$ EXAMPLE: (FOR 1 1/2 FULL BORE) $A = \dfrac{(1.5)^2 \times 3.1416}{4} = 1.767\ in.^2$.

FIGURE 3.7
Cylinder application data.

Solution: (see Table 2.1 for U.S. conversion fractions.) The cylinder speed formula is:

$$\text{Speed} = \text{Flow} \times \frac{1}{\text{Area}}$$

$$= \frac{5 \text{ gal}}{\text{min}} \times \frac{231 \text{ in.}^3}{\text{gal}} \times \frac{1}{3.142 \text{ in.}^2}$$

$$= \textbf{367.6 in./min} \qquad\blacksquare$$

Note: The gallons cancel and $in.^3/in.^2$ = in. (as shown by the strikethroughs). All that is left are the numbers $\dfrac{5 \times 231}{3.142}$ = 367.6 and the dimensions (in./min).

■ **EXAMPLE 3.5**

How fast would the 2-in. cylinder of Example 3.4 retract with the same 5 gal/min flow (assume a standard rod)?

Solution:

$$\text{Speed} = \text{Flow} \times \frac{1}{\text{Area}}$$

$$\text{Speed} = \frac{5 \text{ gal}}{\text{min}} \times \frac{231 \text{ in.}^3}{\text{gal}} \times \frac{1}{2.357 \text{ in.}^2} = \textbf{490 in./min} \qquad\blacksquare$$

In the United States, the pound is the unit weight or force and when we translate to metric or SI units we think of kilograms. I think this is also true in most other parts of the world, particularly in Europe and Canada even though they have otherwise adopted the SI standards. Using the kilogram as the unit of weight or force has been common practice since the days of Napoleon.

Indeed when you look at the old brass weight labeled "1.0 kg" used in a balance scale, it is really 9.807 newtons by the SI standards. This is because in the SI standard, a newton is the force required to accelerate one kilogram of mass at the rate of one meter per second per second. Observe that, in the SI standard, the *kilogram* is no longer weight or force but now *mass*. Therefore, to be academically correct when using SI units, weight and force should be specified in newtons (giving honor to Sir Isaac Newton, the father of physics).

Note: It is important to understand that, in contrast to this academically correct SI standard, it is common practice in the United States and Europe, especially among hydraulic system designers, to use kilograms of force (kg_f) in their calculations instead of newtons.

The English or U.S. system does not have a unit that is equivalent to the SI kilogram (mass). Years back, the slug was promoted but never was really accepted. Mass is thus expressed as pounds per inch per second per second; that is, pounds of weight per unit of acceleration. In this form mass becomes self-explanatory. No great scientist is honored, but confusion is eliminated and we are educated by the simplicity of dimensional analysis.

Note: A basic physics formula states that

$$\text{Force} = \text{Mass} \times \text{Acceleration } (f = ma)$$

$$\text{or } m = \frac{f}{a}$$

$$\text{or } m = \frac{\text{lb}}{\text{in./sec}^2} = \frac{\text{lb} \times \text{sec}^2}{\text{in.}} = \text{Mass}$$

Cylinder Application Examples (SI Units)

■ ### EXAMPLE 3.6

Suppose you want to select a cylinder size that will lift a 500-kg load with a hydraulic pressure of 5000 kPa, and all you have is the data of Figure 3.7 (U.S. units) and Table 2.3, which shows SI to U.S. unit conversions. Use this procedure:

1. Convert the 500-kg load weight to pounds:

$$500 \ \cancel{\text{kg}} \times \frac{2.2 \text{ lb}}{\cancel{\text{kg}}} = \textbf{1100 lb}$$

2. Convert the pressure from kPa to lb/in.^{2}:

$$\overset{50}{\cancel{5000}} \ \cancel{\text{kPa}} \times \frac{14.5 \text{ lb/in.}^2}{\cancel{100} \ \cancel{\text{kPa}}} = \textbf{725 lb/in.}^2$$

$$\text{or } 5000 \ \cancel{\text{kPa}} \times \frac{\text{lb/in.}^2}{6.895 \ \cancel{\text{kPa}}} = \textbf{725 lb/in.}^2$$

Solution: The smallest possible cylinder piston area is:

$$1100 \ \cancel{\text{lb}} \times \frac{\text{in.}^2}{725 \ \cancel{\text{lb}}} = \underline{\textbf{1.52 in.}^2}$$

From the table of Figure 3.7, the smallest cylinder with a piston area larger than 1.52 in.^{2} is the one with a 1½-in. bore and an area of 1.76 in.^{2}. *Normally this is the one you would select.*

Another procedure for handling Example 3.6 follows:

1. Leave the pressure in kPa:

$$= 5000 \text{ kPa}$$

2. Convert kg$_f$ to newtons:

$$500 \ \cancel{\text{kg}_f} \times \frac{9.8066 \text{ N}}{\cancel{\text{kg}_f}} = \textbf{4903 N}$$

3. Solve for cylinder area in m².

Solution: The smallest piston area is:

$$\frac{\cancel{Pa}}{\cancel{N/m^2}} \times 4903 \ \cancel{N} \times \frac{1}{5,000,000 \ \cancel{Pa}} = \mathbf{0.0009806 \ m^2}$$

$$0.0009806 \ \cancel{m^2} \times \frac{1550 \ in.^2}{\cancel{m^2}} = \mathbf{1.52 \ in.^2} \qquad ■$$

3.2 HYDRAULIC MOTORS

Hydraulic motors are rotary actuators. However, the name "rotary actuator" is reserved for a particular type of unit that is limited in rotation to less than 360°. The continuous-rotation-type hydraulic motors are generally divided into three types: (1) gear, (2) vane, and (3) piston motors.

Common Motor Characteristics

Fixed-displacement hydraulic motors produce speed proportional to flow, and torque proportional to differential pressure. Differential pressure or pressure drop is the difference between the inlet pressure and the outlet pressure. In order for the motor to produce torque and demand pressure from the pump, however, there must be some opposition to its rotation. Something must be trying to hold it back. Otherwise, the pressure drop across the motor will be quite low.

Motor Applications (U.S. Units)

As usual, we first look at motor characteristics using U.S. standards. Later in the chapter we cover the subject with SI metric units.

Figure 3.8 shows how torque is determined. Five pounds of weight at a 5-in. radius equals 25 lb-in. of torque. A weight of 5 lb at a 3-in. radius produces 15 lb-in.

FIGURE 3.8
Torque.

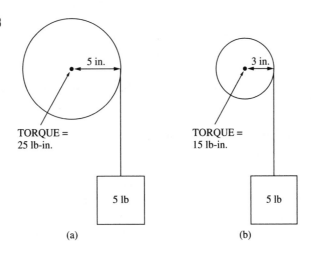

TORQUE =
25 lb-in.

TORQUE =
15 lb-in.

5 lb

5 lb

(a)   (b)

Note: Torque is load force (lb) multiplied by radius (in.). The unit lb-in. contains the same individual dimensions as work "in.-lb." However, they are not the same units; and they never should be combined in a dimensional solution. The best way to handle these units is to label torque as (lb-in.) and work as in.-lb, always treating (lb-in.) as a single dimension.

Motor displacement is the volume of oil that is required for one rotation, or one radian, and is usually expressed as in.^{3}/rev or in.^{3}/rad (there are 2π radians per revolution). With a fixed input flow, the speed of a motor is inversely proportional to its displacement. (The larger the displacement, the slower the speed.)

With a fixed pressure limit to the motor (relief valve or compensated pump), the maximum torque delivered by the motor is directly proportional to its displacement. (The larger the displacement, the greater the torque.)

Motor torque (lb-in.) may be calculated from the pressure drop across the motor (lb/in.^{2}) and the motor displacement (in.^{3}/radian).

Note: Since there are 6.28 radians per revolution, the displacement may be converted from in.^{3}/rev to in.^{3}/rad by multiplying

$$\frac{in.^3}{rev} \text{ by } \frac{rev}{6.28 \text{ rad}} = in.^3/6.28 \text{ rad}$$

■ EXAMPLE 3.7

Determine the output torque of a hydraulic motor with a displacement of 12.56 in.^{3}/rev when the pressure drop across the motor is 1000 lb/in.^{2}. (Assume 100% torsional efficiency.)

Solution: Torque = Pressure × Displacement (in.^{3}/rad)

$$\text{Torque} = \frac{1000 \text{ lb}}{in.^2} \times \frac{\overset{2}{12.56} \; in.^3}{rev} \times \frac{rev}{6.28 \text{ rad}}$$

$$= 2000 \text{ lb-in./rad or just } \textbf{2000 lb-in. torque} \qquad ■$$

Note: There are 2π or 6.28 radians in a revolution or in the circumference of one circle. To obtain in.^{3}/radian we multiplied:

$$\frac{in.^3}{rev} \text{ by } \frac{rev}{6.28 \text{ rad}} = \textbf{in.}^3\textbf{/rad} \text{ (see Table 2.1 in Chapter 2)}$$

Motor Applications (SI Units)

■ EXAMPLE 3.8

(A) Determine the output torque of a hydraulic motor with a displacement of 60 cm^3/rev when the pressure drop across the motor is 100 bars. (Assume 100% torsional efficiency.)
(B) Determine the SI kW power delivered to the load of the motor in part (A) if the speed of the motor is 600 rev/min, and the overall efficiency of the motor is 100%.

Solution A

$$\text{Torque (N-m)} = \frac{\text{Pressure (N/m}^2) \times \text{displacement (m}^3/\text{rev})}{2\pi \text{ rev/rad}}$$

$$(\text{Pressure) N/m}^2 = 100 \text{ bar} \times \frac{100,000 \text{ Pa}}{\text{bar}} \times \frac{\text{N/m}^2}{\text{Pa}} = 10,000,000 \text{ N/m}^2$$

$$\frac{(\text{Displacement) m}^3}{\text{rad}} = \frac{60 \text{ cm}^3}{\text{rev}} \times \frac{\text{rev}}{6.28 \text{ rad}} \times \frac{\text{m}^3}{1,000,000 \text{ cm}^3}$$

$$= \mathbf{0.00000955 \text{ m}^3/\text{rad}}$$

$$\text{Torque (N-m)} = \frac{10,000,000 \text{ N}}{\text{m}^2} \times \frac{0.00000955 \text{ } \frac{\text{m}^3}{\text{m}}}{\text{rad}} = \mathbf{95.5 \text{ N-m}}$$

Solution B.1 Assume 100% overall motor efficiency.

$$\text{Motor output SI power} = \frac{\text{Torque (N-m)} \times \text{Speed (rev/min)}}{9550}$$

$$= \frac{95.5 \times 600}{9550} = \mathbf{6 \text{ kW}}$$

Check this answer by using another SI power formula:

Solution B.2

$$\text{SI Power (kW)} = \frac{(\text{L/min}) \times \text{bar}}{600}$$

$$\text{L/min} = \frac{60 \text{ cm}^3}{\text{rev}} \times \frac{600 \text{ rev}}{\text{min}} \times \frac{\text{L}}{1000 \text{ cm}^3} = \mathbf{36 \text{ L/min}}$$

$$\text{SI Power (kW)} = \frac{36 \text{ (L/min)} \times 100 \text{ (bar)}}{600} = \mathbf{6 \text{ kW}}$$

So, it checks with the first solution, Solution B.1.   ■

Gear Motors

Figure 3.9 shows a typical gear motor. At first glance, you might think that fluid would come in at the bottom, flow up through the center, and out the top. But look again; there is no path through the center. The only way that fluid can move through the motor is to be caught between the teeth of the gears and move around the outer perimeter from bottom to top.

Torque is the result of the hydraulic pressure acting against the area of one tooth. There are two teeth trying to move the rotor in the proper direction, while one net tooth at the center mesh is trying to move it in the opposite direction.

The shaft is connected to one gear while the other one goes along for the ride. The side (or radial) load on the motor bearing is quite great, because all the hydraulic pressure is on one side. This limits the bearing life of the motor.

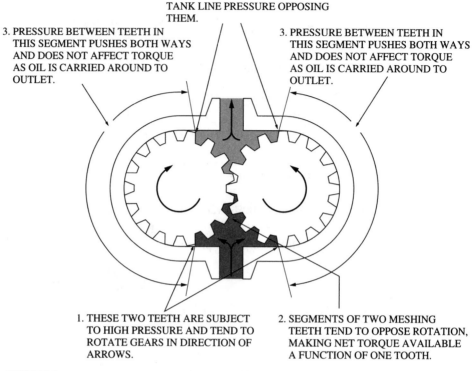

4. THESE TWO TEETH HAVE ONLY TANK LINE PRESSURE OPPOSING THEM.

3. PRESSURE BETWEEN TEETH IN THIS SEGMENT PUSHES BOTH WAYS AND DOES NOT AFFECT TORQUE AS OIL IS CARRIED AROUND TO OUTLET.

3. PRESSURE BETWEEN TEETH IN THIS SEGMENT PUSHES BOTH WAYS AND DOES NOT AFFECT TORQUE AS OIL IS CARRIED AROUND TO OUTLET.

1. THESE TWO TEETH ARE SUBJECT TO HIGH PRESSURE AND TEND TO ROTATE GEARS IN DIRECTION OF ARROWS.

2. SEGMENTS OF TWO MESHING TEETH TEND TO OPPOSE ROTATION, MAKING NET TORQUE AVAILABLE A FUNCTION OF ONE TOOTH.

FIGURE 3.9
Gear motor. (Courtesy of Vickers Inc.)

The gear motor is simple in construction and has good dirt tolerance, but its efficiency is lower than the vane or piston types, and it leaks more than the piston units. Generally, it is not recommended as a servo motor.

Vane Motors

Figure 3.10 shows an unbalanced vane motor, consisting of a circular chamber in which there is an eccentric rotor carrying several spring- or pressure-loaded vanes. Because fluid flowing through the inlet port finds more area of the vanes exposed in the upper half of the motor, the fluid will exert more force on the upper vanes, and the rotor will turn counterclockwise.

The *displacement* of the vane hydraulic motor is a function of its eccentricity.

The radial load on the shaft bearing of the unbalanced vane motor is also large because all of its inlet pressure is on one side of the rotor.

Figure 3.11 shows the answer to the radial bearing load problem of the vane motor. The solution is the double-lobed ring with diametrically opposed ports, which is called a

FIGURE 3.10
Unbalanced vane motor.
(Courtesy of Vickers
Inc.)

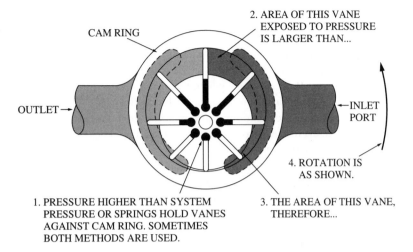

CAM RING

2. AREA OF THIS VANE
EXPOSED TO PRESSURE
IS LARGER THAN...

OUTLET

INLET
PORT

4. ROTATION IS
AS SHOWN.

1. PRESSURE HIGHER THAN SYSTEM
PRESSURE OR SPRINGS HOLD VANES
AGAINST CAM RING. SOMETIMES
BOTH METHODS ARE USED.

3. THE AREA OF THIS VANE,
THEREFORE...

balanced vane motor. In a balanced vane motor, the side force on one side of the bearing is canceled by an equal and opposite force from the diametrically opposite pressure port. The like ports are generally connected internally so that only one inlet and one outlet port are brought outside.

The balanced vane design is a good reliable open-loop motor but has more internal leakage than the piston type and is not generally recommended for servo control systems.

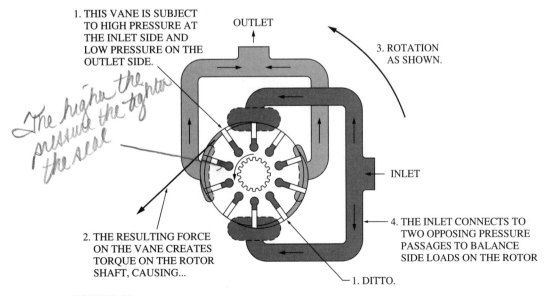

1. THIS VANE IS SUBJECT
TO HIGH PRESSURE AT
THE INLET SIDE AND
LOW PRESSURE ON THE
OUTLET SIDE.

OUTLET

3. ROTATION
AS SHOWN.

The higher the pressure the tighter the seal

INLET

2. THE RESULTING FORCE
ON THE VANE CREATES
TORQUE ON THE ROTOR
SHAFT, CAUSING...

4. THE INLET CONNECTS TO
TWO OPPOSING PRESSURE
PASSAGES TO BALANCE
SIDE LOADS ON THE ROTOR

1. DITTO.

FIGURE 3.11
Balanced vane motor. (Courtesy of Vickers Inc.)

Piston Motors

Piston-type fluid power motors use single-acting pistons that are extended by the fluid pressure and discharge the fluid as they retract. They translate piston motion into circular rotation of the shaft by the use of an eccentric ring, a bent axis, or a swash plate.

Piston motors generally have an odd number of pistons. With this arrangement, although one cylinder may be blocked by the valve crossover, the same number of pistons are receiving fluid as are discharging fluid. With an even number of pistons, and one blocked, there would be one more piston either receiving or discharging fluid, resulting in a pulsation of speed and torque.

Piston motors have the advantage of lower internal leakage than any other type of motor. The reason for this low leakage is the improved seal provided by a long engagement of piston and bore, as well as a tight clearance between piston and block.

Radial Piston Motors

Figure 3.12 shows a radial piston motor. Pistons in the motor work radially against an eccentric track ring to produce rotation. The pintle, which acts as a stationary hydraulic valve, directs fluid to half of the cylinder bores to force the pistons outward radially, while the other side of the pintle allows fluid to be discharged from the retracting pistons. Pistons under pressure can move only by revolving toward the point where the track ring is farthest from the cylinder block axis.

As the pistons rotate, they carry the cylinder block and shaft with them, always keeping the pintle lined up with the proper bores. The pistons not exposed to inlet fluid pressure move toward the cylinder block axis, expelling return oil. The pintle-rotor action is similar to the brush-commutator action of a *dc* electric motor, which provides current to the proper windings to keep the rotor turning.

FIGURE 3.12
Radial piston motor.
(Courtesy of Vickers
Inc.)

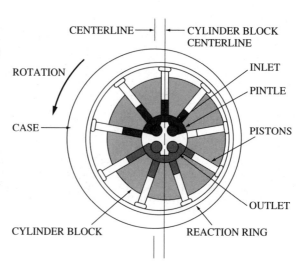

The displacement of a radial piston motor is a function of the area of its pistons and the eccentricity of the track ring. The radial motor is generally limited to higher horsepower sizes.

Axial Piston Motors

The rotor of an axial piston motor rotates on the same axis as the pistons. The two basic designs of axial piston motors are the *bent-axis* and the *in-line* types.

Bent-Axis Piston Motors Figure 3.13 shows a bent-axis piston motor. Oil entering the motor passes through the valve plate to the pistons exposed to the inlet port. These pistons, which are under pressure, push against the shaft flange, which rotates the drive shaft as they move to the point farthest away from the valve plate. The valve plate serves as a commutator, directing oil to the proper pistons to keep the motor turning. *Thus, axial travel of the pistons rotates the cylinder block.*

FIGURE 3.13

Bent-axis piston motor. (Courtesy of Vickers Inc.)

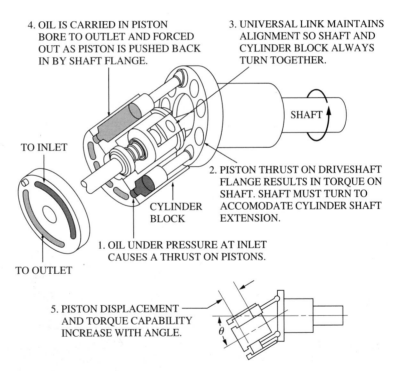

4. OIL IS CARRIED IN PISTON BORE TO OUTLET AND FORCED OUT AS PISTON IS PUSHED BACK IN BY SHAFT FLANGE.

3. UNIVERSAL LINK MAINTAINS ALIGNMENT SO SHAFT AND CYLINDER BLOCK ALWAYS TURN TOGETHER.

SHAFT

TO INLET

2. PISTON THRUST ON DRIVESHAFT FLANGE RESULTS IN TORQUE ON SHAFT. SHAFT MUST TURN TO ACCOMODATE CYLINDER SHAFT EXTENSION.

CYLINDER BLOCK

1. OIL UNDER PRESSURE AT INLET CAUSES A THRUST ON PISTONS.

TO OUTLET

5. PISTON DISPLACEMENT AND TORQUE CAPABILITY INCREASE WITH ANGLE.

θ

Motor displacement is proportional to piston area and number of pistons, and varies as the sine of the angle between the cylinder block axis and the output shaft axis. The maximum practical angle is 30° and the minimum is about 7.5°. With a fixed inlet flow, the smaller the angle the higher the speed; and with a fixed maximum pressure, the smaller the angle the smaller the maximum torque.

The bent-axis piston motor is highly efficient with low leakage. For many years this motor has been used in servo systems, especially aboard aircraft. However, the latest designs for in-line piston motors have virtually replaced the bent-axis type in these applications.

In-Line Piston Motors Figure 3.14 shows an in-line piston motor. The motor drive shaft and cylinder block are centered on the same axis. Hydraulic pressure against the pistons causes a reaction against a stationary canted *swash plate*. With this and the commutator action of the valve plate, the shoe plate, cylinder block, and motor shaft will rotate.

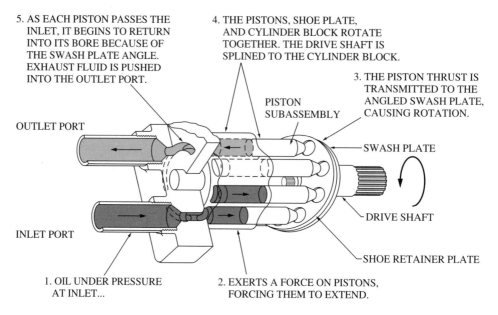

5. AS EACH PISTON PASSES THE INLET, IT BEGINS TO RETURN INTO ITS BORE BECAUSE OF THE SWASH PLATE ANGLE. EXHAUST FLUID IS PUSHED INTO THE OUTLET PORT.

4. THE PISTONS, SHOE PLATE, AND CYLINDER BLOCK ROTATE TOGETHER. THE DRIVE SHAFT IS SPLINED TO THE CYLINDER BLOCK.

3. THE PISTON THRUST IS TRANSMITTED TO THE ANGLED SWASH PLATE, CAUSING ROTATION.

PISTON SUBASSEMBLY

OUTLET PORT

SWASH PLATE

DRIVE SHAFT

INLET PORT

SHOE RETAINER PLATE

1. OIL UNDER PRESSURE AT INLET...

2. EXERTS A FORCE ON PISTONS, FORCING THEM TO EXTEND.

FIGURE 3.14
In-line piston motor. (Courtesy of Vickers Inc.)

The characteristics of the in-line motor are similar to those of the bent-axis unit with the swash plate angle taking the place of the axis angle. The displacement can be varied on either motor by changing this angle.

The big difference in the two axial piston motors is the reduced torsional efficiency of the in-line unit caused by the frictional drag of its shoe plate against its swash plate. Although this extra loss will reduce the output horsepower and can raise the unit temperature slightly, the added friction can increase damping to the servo loop and thus improve the system. The torsional gain loss caused by the drag of the shoe plate is overcome easily by the inherently high pressure gain of a normal servo valve.

Note: Chapter 9 provides more information on servo performance.

■ | **EXAMPLE 3.9**

What would be the speed in revolutions per minute of the motor of Figure 3.13 if the motor displacement is 2 in.^{3}/rev and the flow to the motor is fixed at 5 gal/min?

Solution:

$$\text{Speed} = \text{Flow} \times \frac{1}{\text{Displacement}}$$

$$\text{Speed} = \frac{5 \ \cancel{\text{gal}}}{\text{min}} \times \frac{231 \ \cancel{\text{in.}^3}}{\cancel{\text{gal}}} \times \frac{\text{rev}}{2 \ \cancel{\text{in.}^3}} = \textbf{577.5 rev/min} \qquad \blacksquare$$

Note: Since Displacement = in.³/rev, $\dfrac{1}{\text{Displacement}} = \dfrac{\text{rev}}{\text{in.}^3}$. *The gallons cancel and the in.³ also cancel (as shown by the strikethroughs). All that is left are the numbers (5 × 231)/2 = 577.5) and the dimensions (rev/min).*

3.3 SUMMARY

1. This chapter deals with the output member of the hydraulic system, the cylinder or motor.
2. Cylinders are linear actuators in that the output of a cylinder is straight-line motion.
3. Single-acting cylinders have only one port and exert force in only one direction.
4. Most single-acting cylinders are mounted vertically, and are retracted by the weight of the load.
5. Double-acting cylinders have two power strokes. Fluid may be directed to either end and exhausted from the other.
6. With a fixed flow from the pump, the single-rod differential cylinder will retract faster than it extends.
7. A double-acting cylinder may be used as a single-acting unit by draining the inactive end to tank.
8. The term "rotary actuator" is reserved for a particular type of unit that is limited in rotation to less than 360°.
9. Fixed-displacement hydraulic motors produce speed proportional to flow, and torque proportional to differential pressure.
10. With a fixed hydraulic flow, the speed of a motor is inversely proportional and the torque is directly proportional to displacement.
11. Torque is load force (lb) multiplied by radius (in.), and the unit is lb-in.
12. The gear motor is simple in construction and has good dirt tolerance, but its efficiency is lower than the vane or piston types, and it leaks more than the piston units.
13. The balanced vane motor is a good reliable open-loop motor but has more internal leakage than the piston type and is not generally recommended for servo applications.
14. Piston motors have the advantage of lower internal leakage than any other type of motor, and are recommended for servo applications.

3.4 PROBLEMS AND QUESTIONS:

1. True or false?: Hydraulic cylinders are linear actuators.
2. True or false?: Hydraulic motors are linear actuators.

3. True or false?: Ram-type cylinders are single acting.
4. True or false?: Ram-type cylinders are not practical for long strokes.
5. True or false?: Single-acting cylinders exert force in only one direction.
6. True or false?: Double-acting cylinders have two power strokes.
7. True or false?: The extending and retracting speeds of a differential cylinder are equal.
8. True or false?: A double-acting cylinder may be used as a single-acting unit by draining the inactive end to tank.
9. True or false?: A fixed flow to either end of a differential cylinder will produce equal speeds.
10. True or false?: With a fixed flow from the pump, the single-rod differential cylinder will retract faster than it extends.
11. True or false?: The cylinder piston metal ring seal provides for better servo accuracy.
12. True or false?: The deceleration rate of a cushioned cylinder is determined by the distance of travel.
13. True or false?: Fixed-displacement hydraulic motors produce speed proportional to pressure.
14. True or false?: Fixed-displacement hydraulic motors produce torque proportional to flow.
15. True or false?: In a hydraulic gear motor, torque is the result of the hydraulic pressure acting against the area of two teeth.
16. True or false?: The gear motor leaks less than the piston type.
17. True or false?: The radial load on the shaft bearing of the unbalanced vane motor is large.
18. True or false?: The balanced vane design is a good reliable open-loop motor.
19. True or false?: Piston motors have the advantage of lower internal leakage than any other type of motor.
20. True or false?: The bent-axis piston motor is less efficient than the vane motor.
21. True or false?: The in-line piston motor has less torsional efficiency than the bent-axis type.
22. How fast will a 3¼-in. cylinder extend when supplied with a fixed flow of 10 gal/min? (*Hint:* Use the data of Figure 3.7 to verify your cylinder calculations. Observe that the 3¼-in. cylinder has a full bore area of 8.296 in.². Or to calculate the area without the table: $A = 3.1416 \times D^2/4 = 3.1416 \times (3¼) \times (3¼)/4 = 8.2957875$ in.² Also note that a flow of 10 gal/min = 10 gal/min $\times$ 231 in.³/gal = 2310 in.³/min.)

$$\frac{2310 \, \cancel{\text{in.}^3}}{\text{min}} \times \frac{1}{8.296 \, \cancel{\text{in.}^2}} = 278.45 \text{ in./min}$$

Then recalculate this problem using a 2-in. cylinder.

23. How fast will a 3¼-in. cylinder with a heavy-duty rod retract when supplied with a fixed flow of 10 gal/min? (Hint: Use 5.154 in.² as the annulus area of the 3¼-in. cylinder.)

24. How fast will a 12-cm cylinder extend when supplied with a fixed flow of 20 L/min?

25. How fast will the cylinder of Problem 24 retract if its annular area is half that of its cap end area and it is supplied by the same 20 L/min flow?

26. What is the speed of a hydraulic motor with a displacement of 2 in.^{3}/rev when supplied with a flow of 400 in.^{3}/min?

27. What is the speed of a hydraulic motor with a displacement of 4 in.^{3}/rev when supplied with a flow of 20 gal/min?

28. What is the speed of a hydraulic motor with a displacement of 10 cm^3/rev when supplied with a flow of 30,000 cm^3/min?

29. What is the torque developed by a 3 in.^{3}/rev hydraulic motor when it has a differential pressure of 3000 lb/in.^{2}?

30. What will be the horsepower delivered to the load of the motor of Problem 27 if it has a pressure drop of 1000 lb/in.^{2} and is assumed to have 100% overall efficiency?

4

Directional Control

The direction of flow and movement of the load may be controlled by the use of valves designed for this purpose. Directional valves may be as simple as one-way check valves with only one flow path, or as complex as multiple positioned spool valves with many flow paths.

4.1   CHECK VALVES

Figure 4.1 shows a hydraulic accumulator being charged from a pumping station through a check valve. Fluid is sometimes stored under pressure in an accumulator for later use. After the accumulator has been sufficiently charged, the pump may be stopped and the volume of fluid in the accumulator dispensed to the load through the on–off two-way valve V_2.

Note: The accumulator is defined in Appendix A, "Glossary of Terms," and is covered in more detail in Chapter 8.

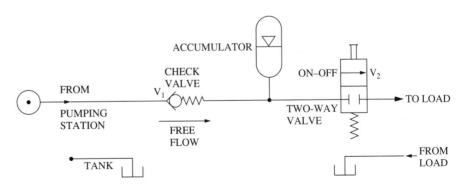

FIGURE 4.1
Accumulator check-valve system.

The check valve, while allowing free flow from the pump, will not permit flow from the accumulator back into the pumping station. *Thus, the check valve's flow is in one direction only.*

■ │ **EXAMPLE 4.1 (U.S. UNITS)**

How long will it take to charge the accumulator of Figure 4.1 to its rated capacity of 150 in.^{3} if the flow from the pumping station is 1.5 gal/min?

Review: We want "time" for the answer, and the only given statement which contains time is 1.5 gal/min. So we write min/1.5 gal to keep time on top where it belongs. Now we must eliminate "gal"; and since we don't have "gal" in any given statement, we must use a conversion fraction, converting gallons to in.^{3} (see Table 2.1).

Solution:

$$\text{Time} = \frac{\text{Time}}{\text{Volume}} \times \text{Volume}$$

$$\text{Time} = \frac{\text{min}}{1.5 \ \cancel{\text{gal}}} \times \frac{\cancel{\text{gal}}}{231 \ \cancel{\text{in.}^3}} \times 150 \ \cancel{\text{in.}^3} = \underline{\textbf{0.4329 min}}$$

$$\text{or } 0.429 \ \cancel{\text{min}} \times \frac{60 \ \text{sec}}{\cancel{\text{min}}} = \underline{\textbf{25.74 sec}} \qquad ■$$

■ │ **EXAMPLE 4.2 (SI UNITS):** How long will it take to charge the accumulator of Figure 4.1 to its rated capacity of 1000 cm^3 if the flow from the pumping station is 16 L/min?

Solution:

$$\text{Time} = \frac{\text{Time}}{\text{Volume}} \times \text{Volume}$$

$$= \frac{\text{min}}{16 \ \cancel{L}} \times \frac{\cancel{L}}{1000 \ \cancel{\text{cm}^3}} \times 1000 \ \cancel{\text{cm}^3} = \underline{0.0625 \ \text{min}}$$

$$\text{or} = 0.0625 \ \cancel{\text{min}} \times \frac{60 \ \text{sec}}{\cancel{\text{min}}}$$

$$= \underline{\textbf{3.75 sec}} \qquad ■$$

Figure 4.2 shows a cutaway view of a ball check valve. Flow entering the left side pushes the ball off its seat and moves through and out to the right. When flow tries to move through the valve from right to left, it finds the ball held on its seat by a spring. Any pressure from right to left will add to the spring force, helping to hold the ball on its seat.

FIGURE 4.2
Ball-spring check valve.

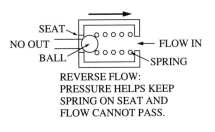

REVERSE FLOW:
PRESSURE HELPS KEEP
SPRING ON SEAT AND
FLOW CANNOT PASS.

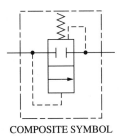

COMPOSITE SYMBOL

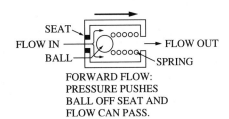

FORWARD FLOW:
PRESSURE PUSHES
BALL OFF SEAT AND
FLOW CAN PASS.

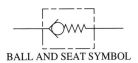

BALL AND SEAT SYMBOL

Graphical symbols for the check valve are shown on the right side of Figure 4.2. The composite symbol shows two movable windows, each depicting a possible flow path. Flow moves from left to right by first acting on the bottom window (dashed lines), moving the bottom window, including the free-flow arrow, up to replace the blocked-ported top window. When flow tries to move from right to left, the top window, with its blocked ports, is held in place (dashed lines), thus preventing flow.

While the composite symbol is sometimes used for the check valve, by and large the most common symbol is the simple "ball and seat" shown at the bottom right of Figure 4.2.

The check valve is the hydraulic equivalent of an electrical diode or rectifier, which allows current to flow in only one direction.

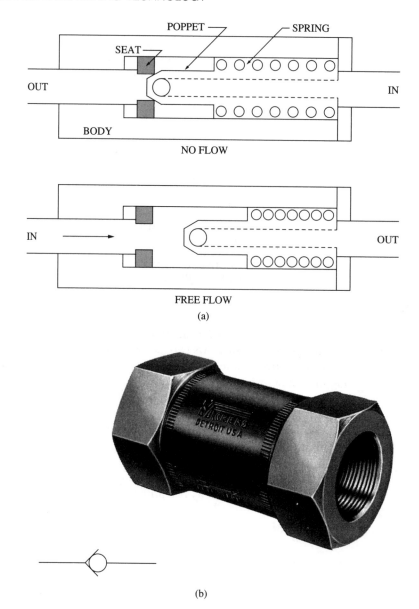

FIGURE 4.3

(a) Cutaway and (b) photograph of an in-line poppet-type check valve. (Courtesy of Vickers Inc.)

In-Line Poppet Check Valve

Figure 4.3 shows a cutaway drawing and a photograph of an in-line check valve using a poppet instead of a ball. The action of the poppet is the same as the ball except that the poppet has a better seating arrangement. The poppet provides a better seal because it always takes the same seat, whereas the ball can rotate and may never seat at exactly the same place twice.

FIGURE 4.4
(a) Cutaway and (b) photograph
of a right-angle check valve.
(Courtesy of Vickers Inc.)

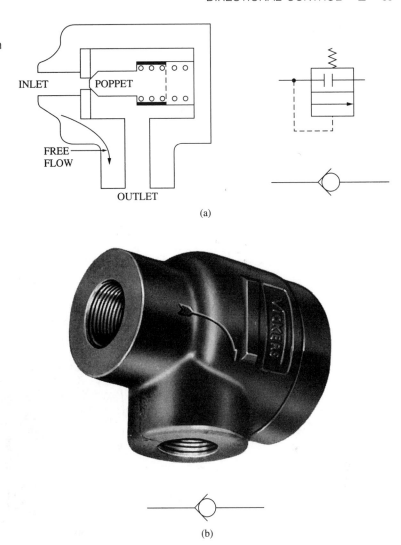

(a)

(b)

The flow through the in-line poppet check valve (from left to right) pushes the poppet off its seat, moves through the hole in the poppet, and out through to the port on the right. Flow trying to move from right to left through the valve will help the spring hold the poppet on its seat, thus preventing flow.

Right-Angle Check Valve

Figure 4.4 is a cutaway drawing and photograph of a right-angle check valve. The action of this valve is similar to that of the in-line valve of Figure 4.3 except that the outlet port is at a right angle to the inlet and the valve is designed for heavier duty use. The flow has less restriction because it does not have to pass through the poppet. The design also lends itself to the use of a heavier spring when needed for higher pressure settings.

FIGURE 4.5

(a) Pilot-operated check valve "to open." (b) Construction of Vickers Model 4C pilot-operated check valve "to open." (Courtesy of Vickers Inc.)

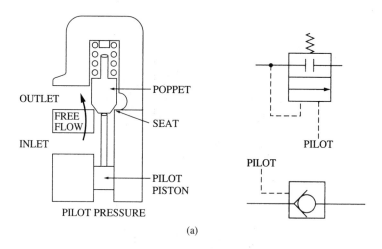

(a)

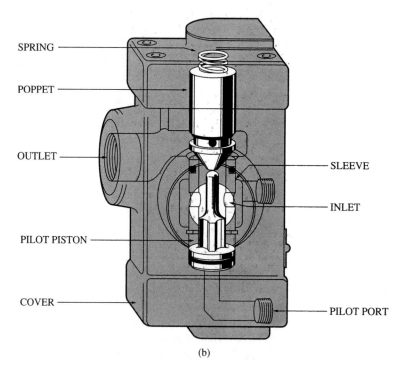

(b)

Pilot-Operated Check Valve

Figure 4.5 shows a cutaway drawing and picture of a pilot-operated (PO) check valve "to open." The action of this valve is the same as the right-angle check except that a piston and plunger are added to push the poppet off its seat with pilot pressure. The valve acts as a regular check until pilot pressure is applied. Then, the poppet is held open, allowing flow to move unrestricted in either direction. This valve is used in a system where it is desirable to stop the check action of the valve for a portion of the machine cycle.

Note: The valve is also available as a pilot-operated check valve "to close," in which the pilot piston is assembled on the opposite end to push the valve closed with pilot pressure.

Care must be given in the selection of the pilot valve to ensure that the pilot piston is large enough to overpower the poppet with the given pilot and poppet pressures.

■ | **EXAMPLE 4.3**

How long will it take for the pilot check valve of Figure 4.5 to open fully for reverse flow if the pilot piston is 5/8 in. in diameter, travels 3/4 in., and is limited to 2.0 gal/min?

Review: We want *time* as the answer, so we start off with min/2 gal. To cancel the "gal" we multiply by gal/231 in.[3]. To cancel 1/in.[3] we must multiply by in.[2] (piston area) and in.[1] (piston travel). To calculate piston area, we use the formula:

$$\text{Area} = \frac{3.1416 \times D^2}{4} = \frac{3.1416 \times 0.625 \text{ in.} \times 0.625 \text{ in.}}{4} = \textbf{0.307 in.}^2$$

Solution: Now we have enough to eliminate all dimensions but time.

$$\text{Time} = \frac{\text{Time}}{\text{Volume}} \times \text{Area} \times \text{Distance}$$

$$= \frac{\text{min}}{2 \text{ gal}} \times \frac{\text{gal}}{231 \text{ in.}^3} \times 0.307 \text{ in.}^2 \times 0.75 \text{ in.} = \textbf{0.000498 min}$$

$$\text{or } 0.000498 \text{ min} \times \frac{60 \text{ sec}}{\text{min}} = \textbf{0.029 sec}$$

$$\text{or } 0.029 \text{ sec} \times \frac{1000 \text{ ms}}{\text{sec}} = \textbf{29 ms} \text{ (milliseconds)}$$

■

4.2 SPOOL-TYPE DIRECTIONAL VALVES

Figure 4.6a shows how a spool is moved inside a valve by a couple of electrically operated solenoids. The lands of the spool open and close discrete ports, providing different flow paths. The spools may be operated manually, mechanically, pneumatically (with air), hydraulically (pilot operated), or electrically.

FIGURE 4.6
(a) Electrically operated valve spool.

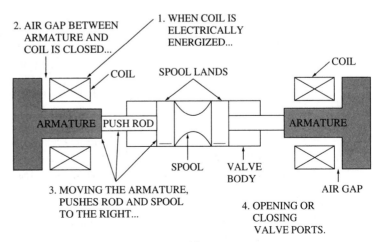

1. WHEN COIL IS ELECTRICALLY ENERGIZED...
2. AIR GAP BETWEEN ARMATURE AND COIL IS CLOSED...
COIL
SPOOL LANDS
COIL
ARMATURE
PUSH ROD
SPOOL
VALVE BODY
ARMATURE
AIR GAP
3. MOVING THE ARMATURE, PUSHES ROD AND SPOOL TO THE RIGHT...
4. OPENING OR CLOSING VALVE PORTS.

(a)

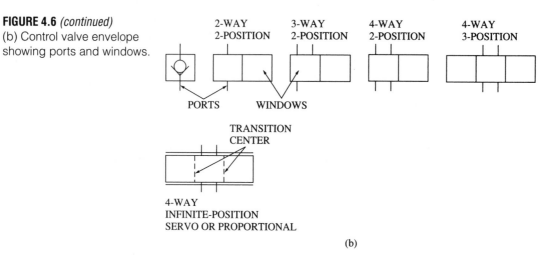

FIGURE 4.6 *(continued)*
(b) Control valve envelope showing ports and windows.

(b)

Figure 4.6b shows the control valve envelopes, windows, and ports of various valve configurations. The internal porting and valve operations will be added as we move along.

Two-Way Valves

Valve V_2 of Figure 4.1 is a two-position, two-way, on–off valve used to connect the accumulator to the load when desired. *A two-way valve is designed to allow flow in either direction between two ports.*

The electrical equivalent of a two-way valve is the regular on–off, single-pole, single-throw (SPST) switch.

Three-Way Valves

Figure 4.7a shows a system where two cylinders are lifted alternately using a three-way valve. Before the pump is started the hand valves V_1 and V_2 are closed. Before the solenoid is energized, the spring holds the spool of the three-way valve to the right position, permitting flow from the pump to pass from P to B, lifting load B all the way to the top.

When the solenoid valve is energized, the spool is moved to the left (the right window is moved to the left), connecting pressure to port A, lifting load A to the top while blocking port B, and keeping load B fully lifted. Hand valve V_2 may now be opened to lower load B as needed.

Before the solenoid is deenergized, valve V_2 should be closed, setting up the next cycle for load B. When the solenoid is deenergized, the spring pushes the spool to the right, bringing the valve porting back to its original setting as shown. Load B is again raised and valve V_1 can now be used to lower load A. Before reenergizing the solenoid, V_1 should be closed to provide for the next cycle.

Figure 4.7b shows a cutaway drawing of a three-way directional valve. View A shows the spool shifted to the left connecting P to A, and the internal spool leakage flow is drained to tank. View B shows the spool shifted to the right, connecting P to B and the spool leakage to tank.

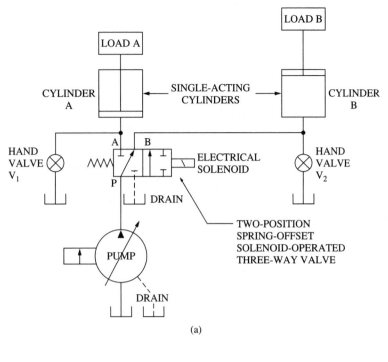

(a)

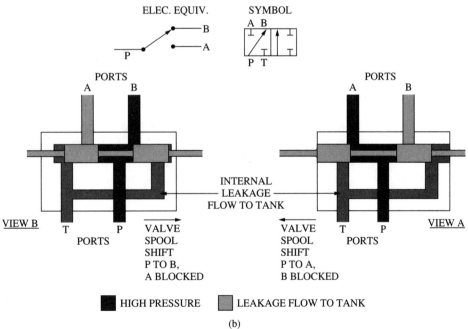

(b)

FIGURE 4.7

(a) System using three-way valve. (b) Three-way valve. Note that this could be considered a four-way valve if the leakage is considered to make port T active.

In the three-way valve of Figure 4.7 there are three active ports: P, A, and B. Port P is connected to A when the solenoid is on, and to B when it is off; this qualifies it as a three-way valve.

The three-way valve is the hydraulic equivalent of an electrical single-pole, double-throw (SPDT) three-way switch.

■ **EXAMPLE 4.4 (U.S. UNITS)**

How long will it take cylinder B of Figure 4.7a to travel from bottom to top if the cylinder has a diameter of $3\frac{1}{4}$ in., a stroke of 20 in., and the pump supplies a flow of 5 gal/min?

Review: We are looking for time in the answer, so we start with min/5 gal. To cancel "gal," we multiply by gal/231 in.^{3}. To eliminate in.^{3} we must know the cylinder area (in.^{2}) and the distance of piston travel (in.). The area of cylinder B is:

$$\text{Area} = \frac{3.1416 \times D^2}{4} = \frac{3.1416 \times 3.25 \text{ in.} \times 3.25 \text{ in.}}{4} = \mathbf{8.29 \text{ in.}^2}$$

We know the travel is 20 in.

Solution:

$$\text{Time} = \frac{\text{Time}}{\text{Volume}} \times \text{Area} \times \text{Distance}$$

$$= \frac{\text{min}}{5 \text{ gal}} \times \frac{\text{gal}}{231 \text{ in.}^3} \times 8.29 \text{ in.}^2 \times 20 \text{ in.} = \mathbf{0.1435 \text{ min}}$$

$$0.1435 \text{ min} \times \frac{60 \text{ sec}}{\text{min}} = \mathbf{8.62 \text{ sec}}$$
■

■ **EXAMPLE 4.5 (SI UNITS)**

How long will it take cylinder A of Figure 4.7a to move from bottom to top if the diameter of the cylinder is 5 cm, the cylinder is 50 cm long, and the pump supplies 15 L/min?

Solution: First solve for the area of the cylinder:

$$\text{Area} = \frac{3.1416 \times D^2}{4} = \frac{3.1416 \times 5 \text{ cm} \times 5 \text{ cm}}{4}$$

$$= \mathbf{19.635 \text{ cm}^2}$$

$$\text{Time} = \frac{\text{Time}}{\text{Volume}} \times \text{Area} \times \text{Distance}$$

$$= \frac{\text{min}}{15 \text{ L}} \times \frac{\text{L}}{1000 \text{ cm}^3} \times 19.635 \text{ cm}^2 \times 50 \text{ cm}$$

$$= \mathbf{.06545 \text{ min}}$$

$$= .06545 \text{ min} \times \frac{60 \text{ sec}}{\text{min}}$$

$$= \mathbf{3.927 \text{ sec}}$$
■

Two-Position, Four-Way Valves

Figure 4.8a shows a system where a double-acting cylinder may be moved in either direction by the use of a two-position four-way valve. As shown in the drawing, the valve

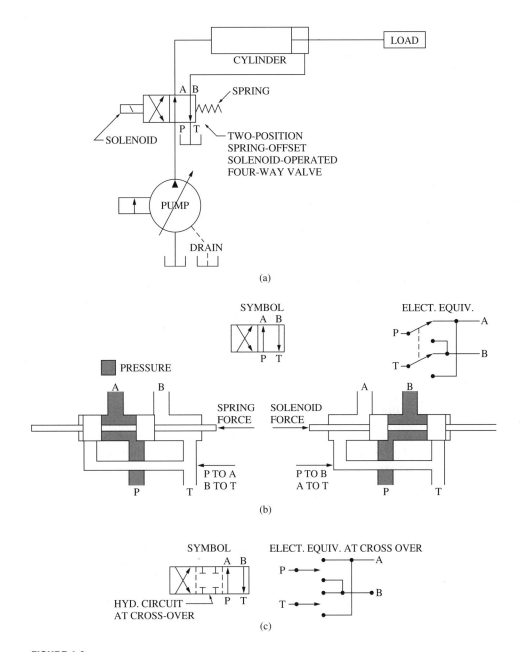

(a)

(b)

(c)

FIGURE 4.8
(a) System using two-position, four-way valve. (b) Two-position, four-way valve cutaway.
(Courtesy of Vickers Inc.) (c) Typical transition symbol.

spool is offset to the left by the spring on the right, porting P to A and B to T. This extends the cylinder, moving the load to the right.

When the electrical solenoid is energized, the spool will shift to the right, placing the left window of the symbol over the ports. This will port P to B and A to T, retracting the cylinder and moving the load to the left.

When the solenoid is deenergized, the spring will shift the spool back to the left, extending the cylinder back to the right.

This valve has no stopped neutral position. The cylinder is either extended or retracted. The only stopped positions of the valve spool are (1) all the way to the right, extending the cylinder, when the solenoid is off; or (2) all the way to the left, retracting the cylinder, when the solenoid is on.

Figure 4.8b shows a cutaway drawing of a two-position four-way valve. There are four active ports in the four-way valve with flow from pressure port P to either A or B; and to tank port T from either A or B. Or:

Flow path from	P to A	when solenoid is off.
Flow path from	B to T	when solenoid is off.
Flow path from	P to B	when solenoid is on.
Flow path from	A to T	when solenoid is on.

The two-position four-way valve is the hydraulic equivalent of an electrical double-pole, double throw (DPDT) switch.

As the two-position four-way valve spool shifts from one position to the other, it passes through an intermediate or center position. If it is essential to the circuit's function to symbolize this "in-transient" condition, it can be shown in a center window of the symbol, enclosed by dashed lines, as illustrated by Figure 4.8c. In this case we are showing a closed center cross-over where all ports (A, B, P, and T) are blocked as the spool passes through center.

Three-Position, Four-Way Valves

Figure 4.9a shows a system using a four-way valve with an actual third (center) position. In this case, energizing solenoid A, while solenoid B is deenergized, connects ports P to A and B to T. Energizing solenoid B, while solenoid A is deenergized, connects ports P to B and A to T. When both solenoids A and B are deenergized, the valve spool returns to center where, in this case, all ports are blocked or closed, thus preventing flow to or from the cylinder.

The cylinder of Figure 4.9a operates much the same as that of Figure 4.8a except that two solenoids are needed and the cylinder can be stopped in any position by deenergizing both solenoids. This allows the springs mounted on both ends of the spool to return it to center.

Caution: When a three-position valve employs two ac type solenoids, the electrical circuit design must provide interlock protection to prevent both solenoids from being energized at the same time. If both solenoids are energized at the same time, they oppose each other, preventing at least one from closing its air gap. If the air gap of either ac solenoid is not closed in a very short time, the solenoid will burn out.

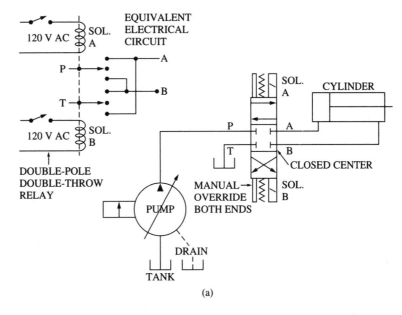

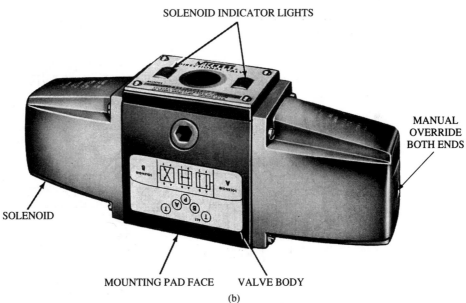

FIGURE 4.9

(a) Three-position, four-way valve system. (b) Vickers Model DG4 S4 valve. (Courtesy of Vickers Inc.)

Figure 4.10 shows six common center conditions for three-position, four-way valves. The valve of Figure 4.9a has a closed center since all ports are closed at center. Valves with the other center conditions shown may be selected for appropriate applications.

FIGURE 4.10

Various center conditions for three-position, four-way valves. (Courtesy of Vickers Inc.)

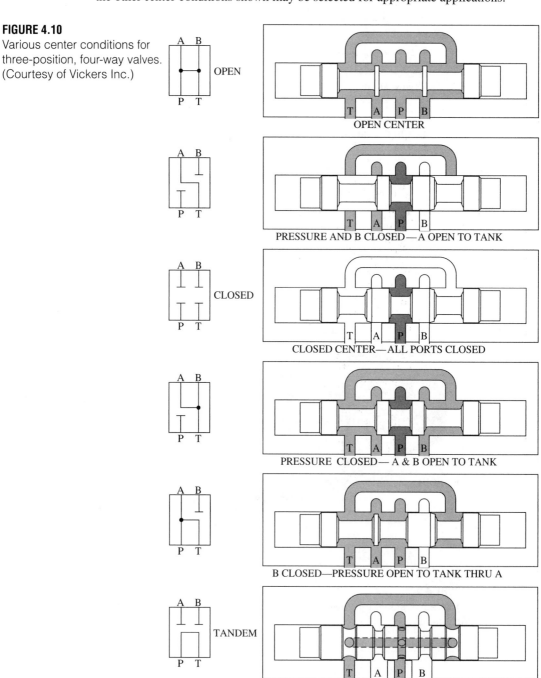

OPEN CENTER

PRESSURE AND B CLOSED—A OPEN TO TANK

CLOSED CENTER—ALL PORTS CLOSED

PRESSURE CLOSED—A & B OPEN TO TANK

B CLOSED—PRESSURE OPEN TO TANK THRU A

TANDEM

The valves of Figure 4.10 all have the same porting arrangements when their spools are shifted to the right. The same is true when they are all shifted to the left. Port A (cylinder port A) is connected to port P (pressure port) and port B (cylinder port B) to T (tank) when the spool is shifted to the left. Port B is connected to P and A to T when the spool is shifted to the right.

Even though the spools of all the valves of Figure 4.10 are constructed differently, the valves function the same except at center. In fact, most manufacturers supply a common valve body in which different spools may be inserted to obtain different center conditions.

The three-position, four-way valve with a closed center is the hydraulic equivalent of an electrical double-pole, double-throw, center off (DPDT, ctr. off) switch or two-coil relay.

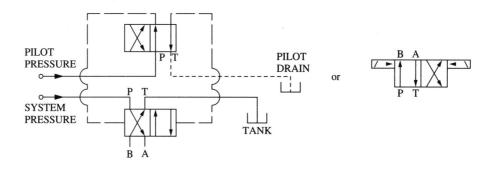

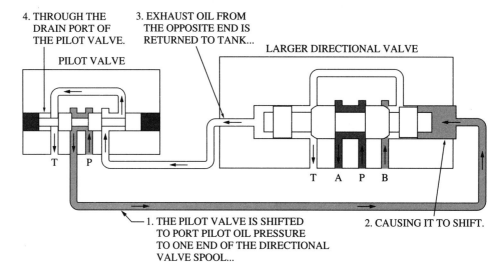

4. THROUGH THE
   DRAIN PORT OF
   THE PILOT VALVE.

3. EXHAUST OIL FROM
   THE OPPOSITE END IS
   RETURNED TO TANK...

LARGER DIRECTIONAL VALVE

PILOT VALVE

1. THE PILOT VALVE IS SHIFTED
   TO PORT PILOT OIL PRESSURE
   TO ONE END OF THE DIRECTIONAL
   VALVE SPOOL...

2. CAUSING IT TO SHIFT.

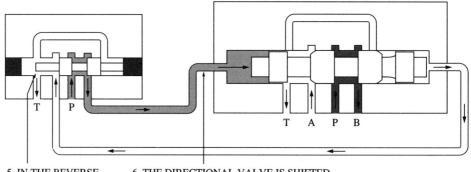

5. IN THE REVERSE
   POSITION OF THE
   PILOT VALVE...

6. THE DIRECTIONAL VALVE IS SHIFTED
   TO THE OPPOSITE POSITION.

FIGURE 4.11
Pilot valves are used to shift larger directional valves. (Courtesy of Vickers Inc.)

Two-Stage, Four-Way Valve

Figure 4.11 shows the internal symbol and hydraulic circuitry for a pilot-operated four-way valve. This valve takes advantage of the large hydraulic force available from the pump to shift the second stage.

The pilot valve of this two-stage arrangement may be operated manually, electrically, pneumatically, or hydraulically. But the larger valve spool is shifted hydraulically by the fluid delivered through the pilot stage.

Figure 4.12, shows the hydraulic circuit and photograph of a Vickers two-stage valve using a pilot choke block and a limited spool travel adjustment. The pilot stage is an electrically operated three-position, four-way valve with a spool type that, when centered, blocks pressure and dumps A and B ports to tank.

Modifications to the second stage (pilot chokes and spool travel limits) are available for applications where shocks to the system, due to valve shifting under pressure, may be objectionable.

FIGURE 4.12

Vickers modified two-stage valve. (Courtesy of Vickers Inc.)

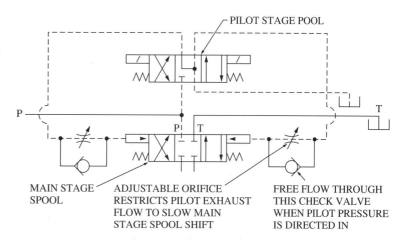

PILOT STAGE POOL

P   T

P   T

MAIN STAGE SPOOL

ADJUSTABLE ORIFICE RESTRICTS PILOT EXHAUST FLOW TO SLOW MAIN STAGE SPOOL SHIFT

FREE FLOW THROUGH THIS CHECK VALVE WHEN PILOT PRESSURE IS DIRECTED IN

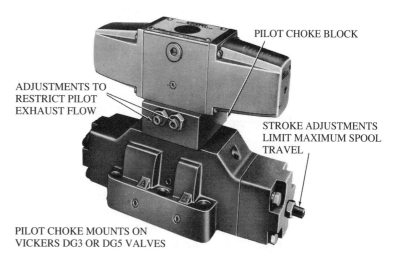

PILOT CHOKE BLOCK

ADJUSTMENTS TO RESTRICT PILOT EXHAUST FLOW

STROKE ADJUSTMENTS LIMIT MAXIMUM SPOOL TRAVEL

PILOT CHOKE MOUNTS ON VICKERS DG3 OR DG5 VALVES

FIGURE 4.13

Two methods of obtaining pilot pressure from non-pressure-compensated fixed-flow pumps. (Courtesy of Vickers Inc.)

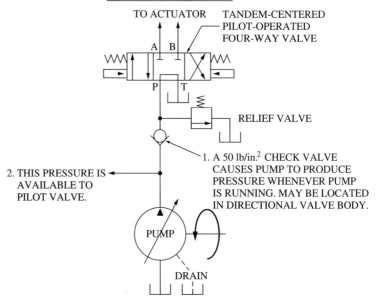

PRESSURE LINE METHOD

TO ACTUATOR   TANDEM-CENTERED
PILOT-OPERATED
FOUR-WAY VALVE

A B

P   T

RELIEF VALVE

1. A 50 lb/in.^{2} CHECK VALVE CAUSES PUMP TO PRODUCE PRESSURE WHENEVER PUMP IS RUNNING. MAY BE LOCATED IN DIRECTIONAL VALVE BODY.

2. THIS PRESSURE IS AVAILABLE TO PILOT VALVE.

PUMP

DRAIN

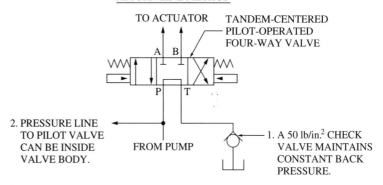

RETURN LINE METHOD

TO ACTUATOR   TANDEM-CENTERED
PILOT-OPERATED
FOUR-WAY VALVE

A B

P   T

2. PRESSURE LINE TO PILOT VALVE CAN BE INSIDE VALVE BODY.

FROM PUMP

1. A 50 lb/in.^{2} CHECK VALVE MAINTAINS CONSTANT BACK PRESSURE.

Pilot Pressure

The two basic types of hydraulic pumps used to supply system pressure are (1) pressure-compensated (fixed-pressure-type) and (2) non-pressure-compensated (fixed-flow-type) pumps.

Both of these types of pumps should be unloaded during standby in order to conserve energy and prevent overheating. To unload the pump, we must reduce its power draw to near zero. Because the power of the pump is the product of *pressure* and *flow*, we must reduce *either* the pressure *or* the flow to near zero at standby.

The pressure-compensated pump automatically strokes back to near-zero flow when the load pressure reaches its compensator setting.

The non-pressure-compensated pump has a constant flow rate, which is fixed by its speed. To unload this pump, we must dump its pressure with an unloading valve, relief valve, or a control valve with an open (P-T) center.

Figure 4.13 shows circuits for two methods that can be used to maintain pilot pressure while the pump pressure has, otherwise, been dumped to tank. The 50 lb/in.^{2} check valve will, in each case, force the pump to operate at 50 lb/in.^{2} even after the relief valve has been dumped or the open centered control valve has been centered. This provides pilot pressure, which will allow the pilot-operated valves to function, thus bringing the system back into operation from standby.

The pressure-compensated pump provides a constant output pressure, which is fixed by the compensator setting, as long as the flow demand of the pump is not more than its rated value. To unload this pump, we merely stop flow demand by returning all control valves to blocked pressure-port conditions during standby. This makes the compensated pump ideal for use with pilot-operated valves because full pressure is available during standby.

Figure 4.14 shows a pressure-compensated pump circuit with pilot pressure taken directly from the pump.

FIGURE 4.14
Pilot pressure from pressure-compensated pump.

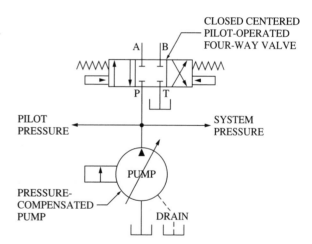

4.3 ## SUMMARY

1. The check valve is a one-way control valve.
2. There is but one flow path in a check valve as shown by the one arrow in its graphical symbol.
3. The check valve is the hydraulic equivalent of an electrical diode, Zener, or rectifier, which allows current to flow in only one direction.
4. Remember that, when comparing electrical switches to hydraulic valves, an open switch is like a closed valve and a closed switch like an open valve.

5. Spool valves may be operated manually, mechanically, pneumatically (with air), hydraulically (pilot operated), or electrically.

6. There are two active ports in a two-way valve with bidirectional flow possible between the two.

7. The three-way valve is the hydraulic equivalent of an electrical single-pole, double-throw (SPDT) switch.

8. There are four active ports in the four-way valve as shown by the four arrows in its graphical symbol.

9. The two-position, four-way valve is the hydraulic equivalent of an electrical double-pole, double-throw (DPDT) switch or a two-coil relay.

10. *Caution:* When a three-position valve employs two ac type solenoids, the electrical circuit design must provide interlock protection to prevent both ac solenoids from being energized at the same time.

11. The three-position, four-way valve with a closed center is the hydraulic equivalent of an electrical double-pole, double-throw, center off switch.

4.4   PROBLEMS AND QUESTIONS

1. True or false?: The check valve is a two-way valve.

2. True or false?: The right-angle check is a heavy-duty valve.

3. True or false?: The ball-spring check valve has a better seating arrangement than the poppet type.

4. True or false?: Flow can be in either direction through a PO "to open" check valve while pilot pressure is applied.

5. True or false?: There are two active ports in a two-way valve.

6. How long will it take the accumulator of Figure 4.1 to charge to its rated capacity of 100 in.^{3} if the pump delivers 1 gal/min?

7. How long will it take the accumulator of Figure 4.1 to charge to its rated capacity of 1600 cm^3 if the pump delivers 4 L/min?

8. How long will it take the PO check valve of Figure 4.5a to open fully, for reverse flow, if the 1/2-in.-diameter pilot piston travels 5/8 in. and the pilot flow is limited to one gal/min?

9. How long will it take the PO check valve of Figure 4.5a to open fully, for reverse flow, if the 1-cm-diameter pilot piston travels 1.5 cm and the pilot flow is limited to 3 L/min?

10. What would happen to the valve spool system of Figure 4.6 if both the ac solenoids were energized at the same time? (Circle one)
    A. The spool would move twice as fast.
    B. The spool would back up.
    C. At least one coil would overheat.
    D. The valve would operate at a lower temperature.

5
Pressure Sensing and Control

Pressure-sensing and control valves are used for the following tasks:

1. Act as a relief valve to limit and regulate system pressure.
2. Unload system pressure to reduce pump heat during standby.
3. Sense load pressure changes to alter flow paths.
4. Counterbalance or brake the load.

5.1 SINGLE-STAGE PRESSURE-SENSING VALVES

Figure 5.1 shows a basic relief valve application. The pump of this example runs at a fixed speed and is not pressure-compensated; therefore its flow is fixed. This type of circuit—where the pump is not pressure-compensated—needs a relief valve to limit and regulate the pressure.

Note: The pressure-compensated pump needs no relief valve. It has a built-in automatic compensator that limits and regulates the upper limit of the outlet pressure.

FIGURE 5.1
Relief valve application.

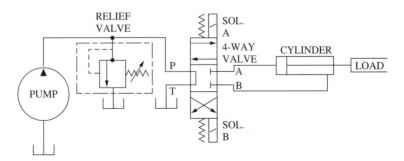

The center porting of the four-way valve (P to T) dumps system pressure during standby. But, if no relief valve were installed, and the cylinder were to stall or bottom out, one or more of the following conditions would occur:

1. The pump would stall.
2. The electric circuit breaker would trip.
3. The pump would relieve itself by rupturing the most vulnerable component of the system.

Simple Single-Stage Direct-Acting Relief Valve

The relief valve can be as simple as a check valve with an adjustable spring. Figure 5.2 shows such a valve. When the full flow of the pump is not needed by the load, the pressure will increase. The increase of pressure will depress the valve spring, pushing the poppet piston off its seat. This will divert the excess flow to tank while maintaining the system pressure near the setting of the relief valve.

The simple single-stage valve of Figure 5.2 has three main disadvantages:

1. It calls for a large spring, which requires a larger sized valve.
2. It does not regulate pressure very well.
3. It has a high pressure override.

FIGURE 5.2
Simple single-stage relief valve.
(Courtesy of Vickers Inc.)

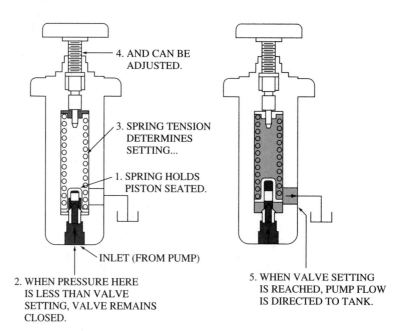

4. AND CAN BE ADJUSTED.

3. SPRING TENSION DETERMINES SETTING...

1. SPRING HOLDS PISTON SEATED.

INLET (FROM PUMP)

2. WHEN PRESSURE HERE IS LESS THAN VALVE SETTING, VALVE REMAINS CLOSED.

5. WHEN VALVE SETTING IS REACHED, PUMP FLOW IS DIRECTED TO TANK.

R-Type Relief Valve

Note: The R valve is a Vickers designation for a series of spool-type valves, which can be changed into a relief, unloading, counterbalance, braking, or sequence valve just by rotating the top or bottom covers. Other manufacturers have similar families of valves. Although the Vickers R valve has given ground to the cartridge valve and other innovations, it is used here as an example of "one valve design performing many functions."

The size problem of the single-stage valve can be reduced by going to an R-type configuration, shown in Figure 5.3. The R-type valve design replaces the poppet with a spool and uses a small piston to move the spool.

The upper range of the valve is extended by applying the sensed pressure to a small piston rather than the larger spool. With the small piston installed, more pressure is needed to move the spool. This increases the pressure setting without increasing the spring size.

The relief valve is designed to regulate and limit the pressure. Neither the poppet nor spool has a fixed position. Either will seek whatever position is necessary, within its capacity as a regulator, to keep the pressure constant.

The reason why the poppet or spool-type single-stage relief valve does not regulate pressure very well is because of its excessive *pressure overide*. Pressure override is the

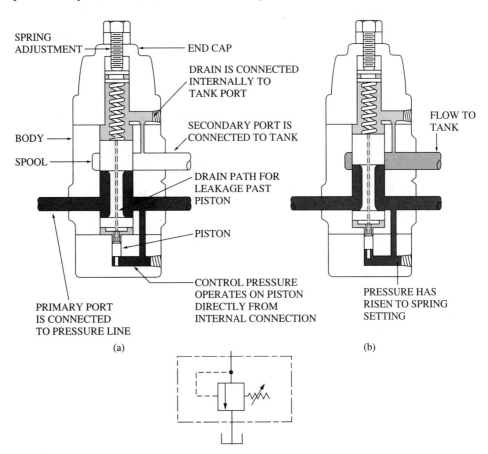

FIGURE 5.3
The Vickers R-type relief valve. (a) Closed. (b) Relieving. (Courtesy of Vickers Inc.)

difference between the cracking pressure—the pressure that just begins to open the valve—and the pressure at full flow through the valve.

If the single-stage relief valve were used just to regulate the pressure—to keep the pressure constant for a wide range of flow requirements—it would do a poor job.

When the flow requirement of the load is reduced, the flow through the relief valve must increase in order to handle the fixed pump delivery. The increased flow through the valve requires further compression of the spring and hence a further increase of pressure from the pump. *Thus, with a single-stage valve, as the flow requirement goes down, the system pressure goes up, losing pressure control.*

Two of the objections to the single-stage relief valve—excessive size and pressure override—can be greatly diminished by going to the two-stage compound pilot-operated relief valve. This valve is covered in detail in Section 5.2. But first, let's explore the many other uses of the R valve.

R-Type Unloading Valve

As stated earlier, the R-type valve is a single-stage, spool-type valve with an adjustable spring pushing against one end of the spool and control pressure against the other end. The control pressure acts against a small piston, which, in turn, moves the spool. The small piston extends the pressure-sensing range of the valve.

The top and bottom covers of the valve are removable and rotatable to new positions, allowing for different control and drainage configurations, making it a very versatile valve indeed. The valve is internally drained to the secondary port when the top cover is in one position and can be externally drained after the cover has been rotated.

Control pressure is taken from the primary port when the bottom cover is in one position and may be externally connected after the cover has been rotated.

Figure 5.4 shows an unloading valve used to dump or "unload" one pressure with another pressure. The bottom cover is rotated so as to allow access of a remote pressure to the control piston. When the remote pressure exceeds the setting of the spring, the spool will be pushed wide open, allowing unrestricted flow to tank.

Here, we do not want to regulate the pressure, but to dump it when the remote pressure indicates such an action is needed. The R-type unloading valve does not meter flow and can be rammed open by the remote control pressure without any detrimental effect caused by the valve's otherwise large pressure override.

The top cover of the unloading valve is in a position that allows the valve to be internally drained to the secondary port. This is acceptable since the secondary port is connected to tank in this application, as it was in the relief valve application.

R- and RC-Type Sequence Valves

Figure 5.5a shows the circuit of two cylinders operated in sequence as controlled by a sequence valve. When solenoid A is energized, the four-way valve will shift to the upper window of the valve symbol, and cylinder 1 will extend until it bottoms out. Cylinder 1 extends first because, when there is a choice of flow paths, the flow takes the path that requires the least pressure.

Note: The RC valve is an R valve with a built-in bypass check to allow free flow in the opposite direction.

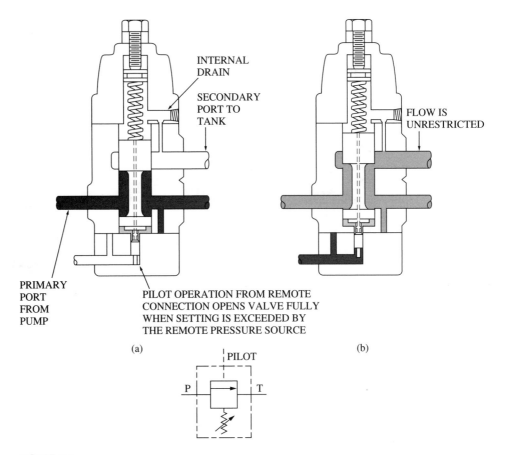

FIGURE 5.4
R-type unloading valve. (a) Closed. (b) Unloading. (Courtesy of Vickers Inc.)

FIGURE 5.5
(a) Sequence valve circuit. (Courtesy of Vickers Inc.)

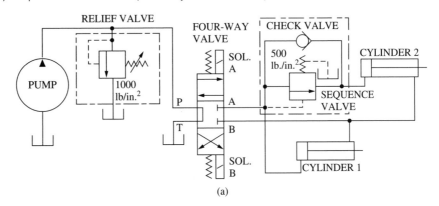

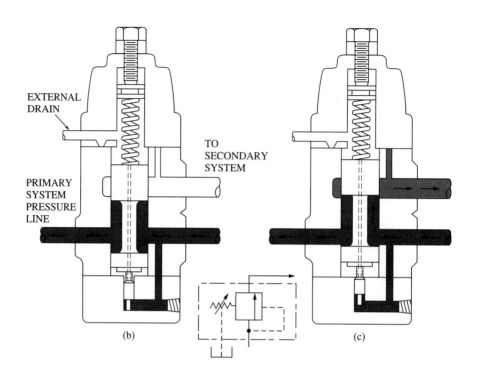

EXTERNAL
DRAIN

TO
SECONDARY
SYSTEM

PRIMARY
SYSTEM
PRESSURE
LINE

(b)

(c)

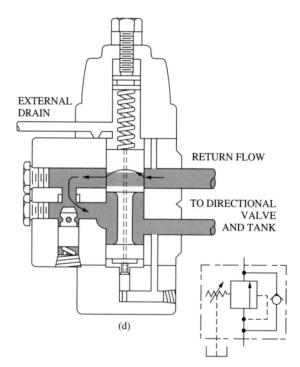

EXTERNAL
DRAIN

RETURN FLOW

TO DIRECTIONAL
VALVE
AND TANK

(d)

FIGURE 5.5 *(continued)*
(b) Closed. (c) Open. (d) RC-type sequence valve, which permits free flow in the reverse direction. (Courtesy of Vickers Inc.)

The pressure of the pump will now build up. When it exceeds the 500 lb/in.^{2} setting of the sequence valve, the valve will open fully to the secondary port. This will extend cylinder 2 until it bottoms out or solenoid A is turned off.

When solenoid B is energized, the four-way valve will shift to the lower window of the valve symbol and both cylinders will retract. The retracting flow from cylinder 2 will bypass the sequence valve through the check valve. The cylinder requiring the least pressure will retract first. If the cylinders must be retracted in a particular sequence, a second sequence valve may be installed.

Figure 5.5b, c, and d shows an R-type valve assembled as a sequence valve. The top cover is rotated to a position that requires an external drain connection. The valve cannot be internally drained to the secondary port because of the high pressure seen at this port for part of its cycle.

The bottom cover is arranged so as to allow control pressure to be taken from the primary system line.

As stated earlier, the RC-type sequence valve is the same as the R-type valve except that it has an integral reverse-flow check valve. It is used in cases where free flow is desired in the reverse direction through the valve, as in the circuit of Figure 5.5a.

RC-Type Counterbalance Valve

Figure 5.6 shows a circuit using an RC-type counterbalance (CB) valve to prevent the cylinder load from free falling due to its heavy weight.

When solenoid B is energized, the bottom window of the four-way valve symbol will be moved up to the center, causing flow from P to B, through the bypass check of the RC valve, into the rod end of the cylinder. The oil from the cap end of the cylinder is returned to tank from A to T of the four-way valve. The cylinder will retract, lifting the load weight. The cylinder, of course, will stop moving when both solenoids are turned off.

When solenoid A is energized, the top window of the symbol will move down to the center; and flow will now move from P to A of the four-way valve, into the cap end

FIGURE 5.6

Counterbalance circuit using an RC-type valve.

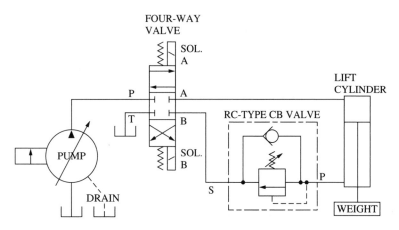

of the cylinder, driving the piston down. Flow out of the rod end of the cylinder will travel through the CB valve from P to S and through the four-way valve from B to T.

Before fluid can move through the RC-type CB valve from P to S, the pressure in the rod end of the cylinder must build up to the spring setting of the CB valve, which is higher than that caused by the hanging weight. This sets up a condition where the load must be forced down, as well as up, by the pump.

If the counterbalance valve were not used, and the rod end of the cylinder returned directly to tank through the four-way valve, the load weight, cylinder rod, and piston would have fallen freely as soon as the four-way valve was shifted to the extension mode.

You might think that the output flow from the rod end of the cylinder would be determined by the input flow to the cap end. This is the condition for most systems; but, in the case of a hanging weight or any other overrunning load, it is not so.

With a hanging weight, the piston will move down faster than it is being pushed by the cap-end flow. This will produce a vacuum above the piston, but the vacuum will not stop the piston from accelerating. A pure vacuum would represent a negative pressure of only -14.7 lb/in.^{2} holding back on the piston. This is not enough to stop a weight of a reasonable size.

The two end covers of the RC valve are set up in the same way for the counterbalance circuit as they are for the R-type relief valve. The control pressure is internally connected to the primary pressure port; and the valve can be internally drained, because the secondary port is connected to tank when the valve is acting as a counterbalance.

RC-Type Motor Brake and Overrun Control Valve

Figure 5.7 shows a circuit using a modified RC valve as a brake for a hydraulic motor as well as speed overrun protection. This is a counterbalance system modified to fit a motor circuit. A motor system is typically different from that of a cylinder system. The motor system usually has more inertia that acts as a flywheel.

The spool of the valve is solid (no drain hole through the center) and a remote control pressure connection is made directly to the bottom end of the spool.

This remote control port is teed into the feed line upstream of the motor. Most of the time we would want to accelerate the motor as fast as possible, with all the system pressure available. In this case, the standard CB valve of Figure 5.6 would be a drag. But if we put the system pressure directly under the full area of the spool, as in the modified valve of Figure 5.7, we will drive the valve wide open, reducing its pressure drop to near zero, during acceleration.

When the motor suddenly reaches its desired speed, it will tend to overspeed due to the inertia (flywheel effect) of its load. This will cause the motor to behave as a pump for a moment, reducing the inlet pressure to the motor and the remote control port of the modified valve. This will allow the valve to revert back to its standard counterbalance configuration. Now the spring of the valve will push the spool down to restrict the flow as in the unmodified version, bringing the speed back under control.

When the four-way valve is centered, flow will be stopped from the pump, but the flywheel effect of the load inertia will drive the motor as a pump until the stored energy of the flywheel is dissipated.

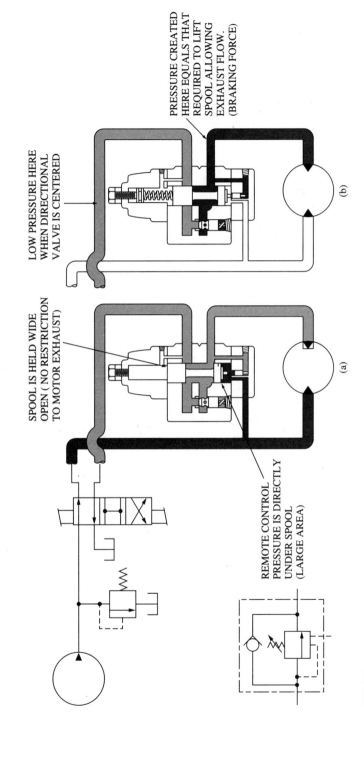

LOW PRESSURE HERE
WHEN DIRECTIONAL
VALVE IS CENTERED

PRESSURE CREATED
HERE EQUALS THAT
REQUIRED TO LIFT
SPOOL ALLOWING
EXHAUST FLOW.
(BRAKING FORCE)

SPOOL IS HELD WIDE
OPEN (NO RESTRICTION
TO MOTOR EXHAUST)

REMOTE CONTROL
PRESSURE IS DIRECTLY
UNDER SPOOL
(LARGE AREA)

(a)

(b)

FIGURE 5.7

Motor brake and speed overrun protection circuit. (a) Acceleration or constant speed. (b) Braking. (Courtesy of Vickers Inc.)

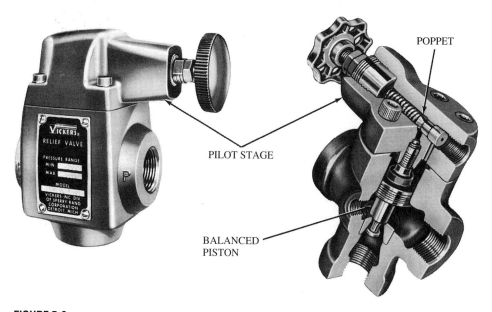

FIGURE 5.8
Two-stage, balanced-piston, compound relief valve. (Courtesy of Vickers Inc.)

As a pump, the motor will draw fluid from the tank, through the four-way valve (open center) and discharge it through the brake valve and the open center of the four-way valve. The discharge pressure of the motor will be whatever is needed to compress the adjustable spring of the brake valve enough to meter the proper flow through the valve.

5.2   TWO-STAGE, BALANCED-PISTON, PRESSURE-SENSING VALVES

Figure 5.8 shows a picture and a cutaway view of a two-stage relief valve. This valve is in common use wherever pressure regulation better than that of a single-stage valve is required.

Adding the second stage to a relief valve is like adding a second transistor to an electronic voltage regulator to reduce its voltage override. In both cases they become better regulators.

The two-stage relief valve has a pilot stage mounted on top of a power stage. The main ports, "pressure" and "tank," are on the power stage; whereas the poppet and spring as well as the remote "vent" connection are on the pilot section. The two stages are joined by a balanced piston.

Figure 5.9 shows how the two-stage balanced-piston valve works. The lower land of the piston acts like a stopper, preventing flow from supply pressure to tank until the piston is lifted. The pressure at the inlet port, acting underneath the large land of the piston, is sensed also on top of the large land through an orifice drilled up through the large land. This hydraulically balances the piston until the pressure reaches the setting of the poppet spring.

When pressure reaches the setting of the spring, the poppet is pushed off its seat, allowing flow through the orifice in the piston, over the poppet, and down the passageway drilled through the center of the piston, to tank.

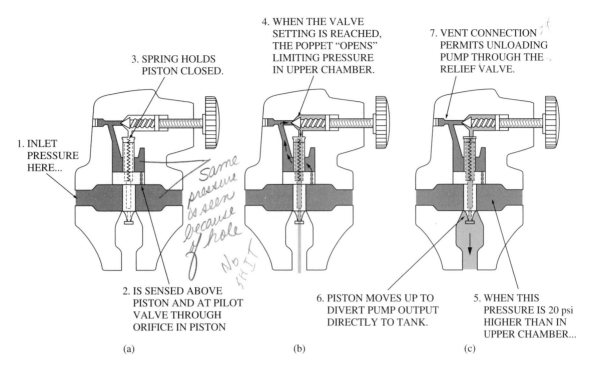

3. SPRING HOLDS PISTON CLOSED.

4. WHEN THE VALVE SETTING IS REACHED, THE POPPET "OPENS" LIMITING PRESSURE IN UPPER CHAMBER.

7. VENT CONNECTION PERMITS UNLOADING PUMP THROUGH THE RELIEF VALVE.

1. INLET PRESSURE HERE...

2. IS SENSED ABOVE PISTON AND AT PILOT VALVE THROUGH ORIFICE IN PISTON

6. PISTON MOVES UP TO DIVERT PUMP OUTPUT DIRECTLY TO TANK.

5. WHEN THIS PRESSURE IS 20 psi HIGHER THAN IN UPPER CHAMBER...

(a)   (b)   (c)

FIGURE 5.9
Operation of a two-stage relief valve. (a) Closed. (b) Cracked. (c) Relieving. (Courtesy of Vickers Inc.)

Note: An orifice is to hydraulics as a resistor is to electronics. There is pressure drop across an orifice when fluid flows through it, just as there is voltage drop across a resistor when current flows through it. ($E = I/R$, where E = voltage, I = current, and R = resistance.)

Flow through the orifice in the large land of the piston creates a pressure drop across the orifice. This results in a higher pressure in the lower chamber, pushing up on the large land of the piston. The piston will be unseated when the pressure drop across the orifice becomes great enough to start compressing the small spring at the top of the piston (20 lb/in.^{2}). This will allow flow from pressure to tank, through the lower part of the valve, at the pressure setting of the poppet spring (which is 20 lb/in.^{2} higher than that needed to crack the poppet off its seat).

As with the single-stage relief valve, when less flow is required by the load of a constant flow pumping system, more flow must be bypassed to tank by the relief valve. This will require more pressure to compress the small spring at the top of the piston—but not much more, because the spring has such a small rate. *So the pressure override of a two-stage relief valve is small compared to a single-stage valve, thus providing much better pressure regulation.*

Remote Venting of Relief Valve

Figure 5.10 shows a circuit for venting a relief valve through a two-position, four-way valve, which has been modified to act as an on–off valve. Because ports P and A of the

FIGURE 5.10

Remote venting of relief valve.
(Courtesy of Vickers Inc.)

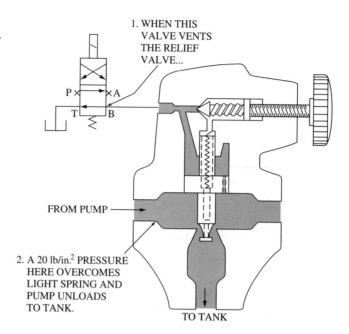

1. WHEN THIS
VALVE VENTS
THE RELIEF
VALVE...

2. A 20 lb/in.² PRESSURE
HERE OVERCOMES
LIGHT SPRING AND
PUMP UNLOADS
TO TANK.

FROM PUMP

TO TANK

four-way valve are blocked, there is but one path through the valve (from B to T), and this is with the solenoid turned off. So, with the solenoid off, the upper end of the balanced piston is ported to tank, limiting the pump pressure to slightly more than 20 lb/in.², the amount needed to compress fully the spring above the piston.

This circuit provides a neat way to start the pump without a load. Just leave the solenoid turned off until the pump is started. With the proper electric circuit for the solenoid, this can also be a fail-safe system. With a momentary interruption of electric power, the solenoid would drop out with the pump, but would not come back on until the solenoid circuit is reset.

FIGURE 5.11

Remote control of relief valve.
(Courtesy of Vickers Inc.)

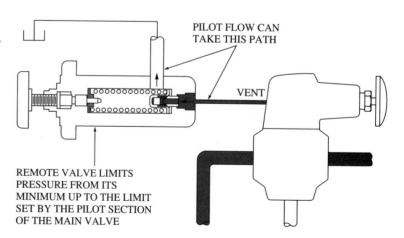

PILOT FLOW CAN
TAKE THIS PATH

VENT

REMOTE VALVE LIMITS
PRESSURE FROM ITS
MINIMUM UP TO THE LIMIT
SET BY THE PILOT SECTION
OF THE MAIN VALVE

Remote Control of Relief Valve

Figure 5.11 shows a circuit using a single-stage relief to control a two-stage relief valve remotely. The pressure port of the single-stage valve is connected to the vent port of the two-stage valve and its secondary port is connected to tank. This parallels the single-stage relief valve with the poppet of the two-stage balanced valve.

To set up this system, adjust the poppet spring of the two-stage valve to the very highest setting allowed by the system, and use the adjustment of the single-stage valve to set the desired system pressures.

Some pressure override will occur across the single-stage valve, but pressure override is a function of flow past the poppet. Also, since very little flow is needed to vent the two-stage valve, the pressure override will be quite small.

Two-Stage Sequence Valve

Figure 5.12 shows a compound two-stage sequence valve. It is like a two-stage balanced-piston relief valve except that it has an external drain port and the drain passage through the piston stem has been eliminated.

The two-stage sequence valve will operate as a relief valve if so connected. But it is a good sequence valve when connected accordingly. The valve of Figure 5.12 could replace the sequence valve of Figure 5.5a. Also, since the two-stage valve has less pressure override, it will provide a snappier transition from primary to secondary control.

The two-stage sequence valve of Figure 5.12 does not need a reverse-flow check valve if the system will allow a 20 lb/in.^{2} back pressure across the valve. A pressure of 20 lb/in.^{2} (plus) will push the piston off its seat and allow free flow from the secondary port back to the primary port (just like a check valve).

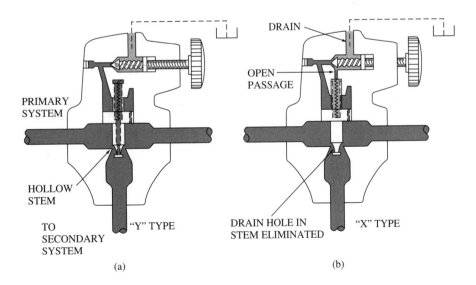

(a)   (b)

FIGURE 5.12
Two-stage sequence valves. (Courtesy of Vickers Inc.)

5.3   PRESSURE-REDUCING VALVE

The pressure-reducing valve is designed to lower and regulate the pressure downstream of the valve. The valve is normally an open circuit, from pressure port to load port, until the load pressure builds up to the setting of the valve spring. At this pressure, the valve partially closes and will maintain this reduced pressure at a constant value as the load restriction continues to increase.

Direct-Acting Pressure-Reducing Valve

Figure 5.13a shows a direct-acting pressure-reducing valve. The reducing valve contains a movable spool with an adjustable spring acting on one end and a pressure-sensing port at the other. The pressure-sensing port is internally connected to the "outlet" port. A "bleed" passage (orifice) is drilled lengthwise through the spool, and the spring end of the spool is ported to tank (external drain).

FIGURE 5.13

Direct-acting pressure-reducing valve. (a) Below valve setting. (b) At valve setting. (c) Circuit using a pressure-reducing valve. (Courtesy of Vickers Inc.)

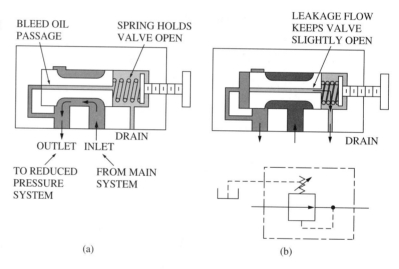

(a)

(b)

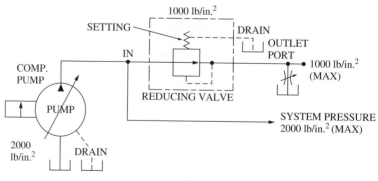

(c)

As stated, the valve remains open, with very little pressure drop from inlet to outlet, until the load pressure builds. Any increase of load pressure will be sensed by the control end of the spool and a small flow will take place through the orifice of the spool and out to the drain.

When the pressure at the outlet port of the reducing valve approaches the reduced-pressure setting of the valve spring, the spool will be pushed to the right, increasing the restriction between the inlet and outlet ports. This increased restriction will hold the outlet pressure constant, even with a reduced flow demand from the reduced-pressure system.

Figure 5.13c shows a reducing valve application. The pressure at the outlet port of the reducing valve remains constant at 1000 lb/in.^{2}, while the total valve drop increases, to keep the pump pressure at 2000 lb/in.^{2}. At this compensator setting, the system pressure will remain at 2000 lb/in.^{2} and the reduced pressure will remain at 1000 lb/in.^{2}, even when the flow of the reduced pressure circuit is reduced further.

Pilot-Operated Pressure-Reducing Valve

As with the balanced-piston relief valve, the pilot-operated reducing valve (Figure 5.14) has a balanced spool where the balance may be upset by a poppet and spring setting. A small spring at the top of the spool holds the spool open until the reduced pressure approaches the valve setting. At this time, the spool will move up to the control position where it will maintain the outlet (control) pressure at the setting of the poppet spring.

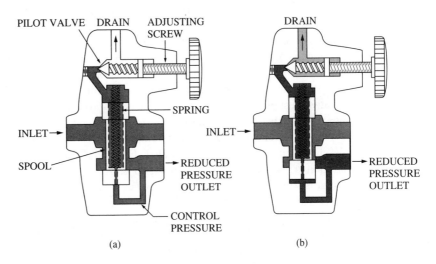

FIGURE 5.14
Pilot-operated pressure-reducing valve. (a) System pressure is below valve setting.
(b) Regulating secondary system pressure. (Courtesy of Vickers Inc.)

FIGURE 5.15
Pressure-reducing valve with
internal check valve.
(Courtesy of Vickers Inc.)

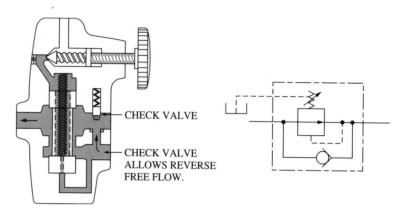

An external drain is required, as with the direct-acting reducing valve, because no other ports are returned to tank.

The pilot-operated valve functions much the same as the direct-acting valve except that it provides more accurate control of the reduced pressure. It also provides for a wider range of pressure control. Figure 5.15 is a pressure-reducing valve with an internal reverse-flow check valve.

Check Valves Compared to Zener Diodes

Zener diodes are sometimes used in electrical circuits to obtain lower and regulated voltages, similar to pressure-reducing valves in hydraulic circuits.

The Zener is a special diode used to provide a reduced regulated voltage. It is a regular diode (check valve) in the forward direction, but breaks down and conducts in the reverse direction at a predetermined design voltage. The Zener is a good analog for a check valve and the standard reducing valve. The Zener can be selected for its breakdown voltage, just as the check valve may be picked for its cracking pressure.

Figures 5.16 and 5.17 show check valves and Zeners in parallel and series arrangements to illustrate a point of similarity. A version of these electrical examples is found quite often, but the hydraulic equivalents may not be as common unless you have an innovative designer aboard. The circuits can be proved, however, if you have an opportunity to build them in a laboratory.

Figure 5.16a shows a pressure-compensated pump feeding through a restriction to three check valves connected in parallel. Each check has a shutoff valve in series. The restriction is provided to allow the pump to keep its 100 $lb/in.^2$ compensator setting while the load is supplied with a selection of lower pressures.

Remember that, when comparing electric switches to hydraulic valves, an open switch is like a closed valve and a closed switch like an open valve. This is not a breakdown of the electrical analogy, but a case of American English usage. We say that a valve is open allowing flow to take place, but we say that an electric switch is open preventing current from passing.

When any valve (V_1, V_2, or V_3) is opened by itself, the load pressure P_L will change to the rated pressure of the corresponding check valve (see the conditions listed in Figure 5.16a).

FIGURE 5.16
(a) Check valves in parallel
compared with (b) Zener diodes
in parallel.

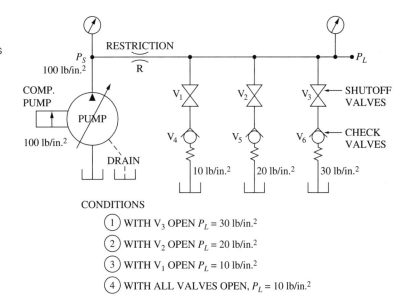

(a)

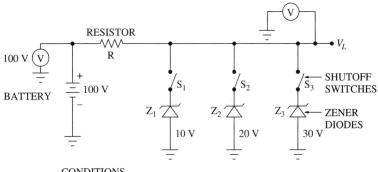

(b)

Consider condition 4 of Figure 5.16a. When valve V_1 is open, check valve V_4 conducts flow at a 10 lb/in.^{2} pressure drop. This causes the load pressure P_L to go to and stay at 10 lb/in.^{2}, leaving 90 lb/in.^{2} across the restriction. Neither of the other check valves will be able to conduct flow, because their pressures are also limited to 10 lb/in.^{2} (not enough to compress the 20 or 30 lb/in.^{2} springs).

Figure 5.16b shows a fixed voltage source (battery) feeding through a resistor to three Zener diodes arranged in a parallel circuit. A disconnect switch is provided for each

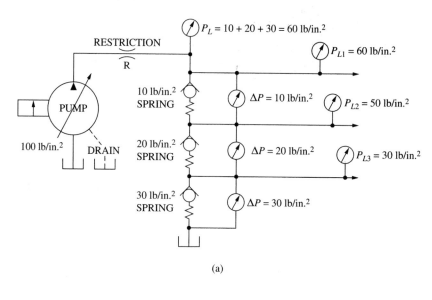

(a)

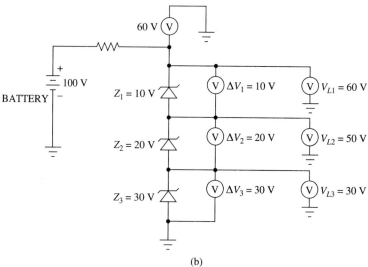

(b)

FIGURE 5.17
(a) Check valves in series compared to (b) Zener diodes in series.

Zener. The resistor is added to allow the battery voltage to remain at 100 volts while the load voltage is dropped to the voltage rating of the active Zener.

When either switch is closed, by itself, the load voltage V_L is changed to the rated voltage of that particular Zener. Consider condition 1: The load voltage V_L will drop to and hold constant at 30 V. Now consider condition 4: The load voltage V_L will drop to and hold constant at 10 V.

Note: When switch S_1 is closed, the load voltage drops to and holds at 10 V. This prevents either of the other Zeners from conducting. So, when any other switch is closed, placing two or more Zeners in parallel, the one with the lowest voltage prevails.

If, as you analyze and compare Figures 5.16a and 5.16b, you could interchange the following terms:

Hydraulics	**Electrics**
Pump	Battery
Pressure	Voltage
Restriction orifice	Resistor
Check valve or single-stage relief valve	Zener diode
Valve open	Switch closed
Valve closed	Switch open

You could read either description for either figure.

Figure 5.17 shows check valves compared with Zener diodes in series circuits. Figure 5.17a shows three check valves with different operating pressures: 10, 20, and 30 lb/in.^{2}. The sum of all pressure drops across the check valves plus the drop across the restriction must add to the pump supply pressure:

40 lb/in.^{2} + 10 lb/in.^{2} + 20 lb/in.^{2} + 30 lb/in.^{2} = **100 lb/in.^{2}**

The three load pressures, P_{L1}, P_{L2}, and P_{L3}, may be used as separate pressure supplies as long as the total load flow does not exceed the capacity of the input pump and restriction.

Figure 5.17b shows the electrical equivalent of the hydraulic circuit of Figure 5.17a. Just use the electrical counterpart for each hydraulic term of Figure 5.17a and you will arrive at Figure 5.17b.

5.4   SUMMARY

1. If some means were not used to limit the output pressure of a constant delivery pump—whenever the pump delivery is shut off—either:
   A. The pump would stall.
   B. The electric circuit breaker would trip.
   C. The pump would relieve itself by rupturing the most vulnerable component of the system.
2. The pressure-compensated pump needs no relief valve. It has a built-in automatic compensator that limits and regulates the outlet pressure to a controlled setting, by reducing the flow delivery of the pump, thus eliminating wasted horsepower, which creates heat in the system.
3. The relief valve is designed to regulate as well as limit the pressure.
4. The single-stage valve has three main disadvantages: (1) It calls for a large spring, which requires a larger sized valve; (2) it does not regulate pressure very well, and (3) it has a high pressure override.

5. The single-stage relief valve (poppet or spool type) does not regulate pressure very well because of its excessive "pressure override."

6. Pressure override is the difference between the cracking pressure—the pressure that just begins to open the valve—and the pressure at full flow through the valve.

7. With a single-stage relief valve, as the flow requirement goes down, the system pressure goes up; and the valve loses pressure control.

8. The two objections of the single-stage relief valve—excessive size and pressure override—can be greatly reduced by going to the two-stage compound relief valve.

9. The R-type valve replaces the poppet with a spool and uses a small piston to move the spool.

10. The top and bottom covers of the R valve are removable and rotatable to new positions, allowing for different control and drainage configurations, which makes it a very versatile valve indeed.

11. The R-type unloading valve does not meter flow and can be rammed open by the remote pressure without any detrimental effect caused by the valve's large pressure override.

12. The RC-type sequence valve is the same as the R-type valve except that it has an integral reverse-flow check valve.

13. The RC-type counterbalance valve is used to prevent the cylinder load from free falling due to its heavy weight.

14. The motor system usually has a hunk of mass some place that acts as a flywheel. A cylinder system moving a large steel table has a mass that also acts like a flywheel.

15. Adding the second stage to a relief valve is like adding a second transistor to an electronic voltage regulator to reduce its voltage override. In both cases, they become better regulators.

16. Flow through the orifice in the piston of a two-stage relief valve creates a pressure drop across the orifice. This results in a higher pressure in the lower chamber, unseating the valve.

17. An orifice is to hydraulics as a resistor is to electronics. There is pressure drop across an orifice when fluid flows through it, just as there is voltage drop across a resistor when current flows through it. ($E = I/R$, where E = voltage, I = current, and R = resistance.)

18. The pressure override of a two-stage relief valve is small compared to a single-stage valve, thus providing much better pressure regulation.

19. The pressure-reducing valve is designed to lower and regulate the pressure downstream of the valve.

20. The pilot-operated pressure-reducing valve functions in the same way as the direct-acting valve except that it more accurately controls the reduced pressure. It also provides for a wider range of pressure control.

21. The electric Zener diode is a good analog of a check valve and restriction used as a pressure-reducing valve.

22. When comparing electric switches to hydraulic valves, an open switch is like a closed valve and a closed switch is like an open valve.

5.5   PROBLEMS AND QUESTIONS

1A. What is the cracking pressure of the relief valve of Figure 5.1 if the area of the poppet piston is 0.5 in.^{2} and the spring is adjusted to a force of 35 lb?

1B. What is the cracking pressure of the relief valve of Figure 5.1 if the area of the poppet piston is 1 cm^2 and the spring is adjusted to a force of 50 kg? [*Hint:* 50 kg × (9.80665 N)/kg = **490.33 N** (see Table 2.2).]

2A. How far will the poppet of Problem 1A be opened when the pressure is adjusted to 90 lb/in.^{2} if the rate of the poppet spring is 5 lb/in.?

2B. How far will the poppet of Problem 1A be opened when the pressure is adjusted to 500 kPa if the poppet spring rate is 2 kg/cm?

3. What is the percent regulation of the valve of Figure 5.1 if the cracking pressure is 60 lb/in.^{2} and the pressure at full flow is 80 lb/in.^{2}? [*Hint:* Percent regulation is equal to the change divided by the original setting. Or (80 - 60)/60 = 20/60 = 0.33 = 33%. Now, redo Problem 3 using a 30 lb/in.^{2} pressure override with the same valve setting (60 lb/in.^{2}).]

4A. How long will it take cylinder 1 of Figure 5.5a to extend fully if it has a length of 12 in., a bore diameter of 3.25 in., and the flow from the pump is 3.5 gal/min? [*Hint:* Piston area = 3.1416 × (3.25 in.)^2/4.]

4B. Rework Problem 4A using the following dimensions: length of cylinder 1 = 30 cm; bore diameter of cylinder 1 = 8 cm; and flow from pump = 10 L/min.

5A. How long will it take cylinder 1 of Problem 4A to retract fully if the rod diameter is 2 in.? [*Hint:* Rod area = 3.1416 × (2 in.)^2/4.]

5B. How long will it take cylinder 1 of Problem 4B to retract fully if the rod diameter is 5 cm? [*Hint:* Rod area = 3.1416 × (5 cm)^2/4.]

6A. What will be the flow from the cap end of cylinder 1 of Figure 5.5A given the information of Problems 4A and 5A?

6B. What will be the flow from the cap end of cylinder 1 of Figure 5.5A given the information of Problems 4B and 5B?

7A. What will be the lifting speed (in./min) of the weight in Figure 5.6 if the cylinder has a 4-in. bore, a 2-in. rod, and the pump flow is 10 gal/min?

7B. Rework Problem 7A using the following dimensions: bore diameter of cylinder = 10 cm; rod diameter of cylinder = 5 cm; and pump flow = 40 L/min.

8. True or false?: A two-stage relief valve has more pressure differential than a single-stage type.

9. True or false?: A two-stage relief valve has better pressure regulation than a single-stage type.

10. The R-type valve replaces the poppet with a _____ and uses a small _____ to move the spool.
Select the proper letter for each blank above.
(A) spring   (B) piston   (C) check   (D) spool.

11. True or false?: The spring-size problem of a single-stage relief valve is solved in the R-type design.

12. True or false?: The R-type unloading valve can be rammed open, but with a resulting detrimental effect of pressure override.
13. True or false?: The R-type valve is a two-stage balanced-piston relief valve.
14. The R-type valve is internally drained to the _____ port when the _____ cover is in one position and can be _____ drained after the cover has been rotated.
    Select the proper letter for each blank above.
    (A) externally (B) secondary (C) primary (D) top (E) bottom.
15. True or false?: The bottom cover of the R-type sequence valve is arranged so as to allow control pressure to be taken from the primary system line.
16. True or false?: The two end covers of the RC valve are set up in the same way for the counterbalance circuit as they are for the R-type relief valve.
17. True or false?: A motor system requires the same counterbalance valve as a cylinder system.
18. In a balanced piston relief valve the _____ will be unseated when the _____ drop across the orifice becomes great enough to start compressing the small spring at the top of the _____ (20 lb/in.^{2}).
    Select the proper letter for each blank above.
    (A) spring (B) pressure (C) piston
19. True or false?: The pressure override of a two-stage relief valve is small compared to a single-stage valve.
20. True or false?: The pressure-reducing valve is designed to lower and regulate the pressure downstream of the valve.
21. True or false?: A closed switch is analogous to a closed valve.

6
Flow Control

The control of flow is a very important consideration for the designer and user of hydraulics. Flow and displacement establish the speed of a cylinder or motor, and the acceleration or deceleration of the load is determined by the rate of flow change.

6.1 FLOW VERSUS RESTRICTION

Occasionally flow is regulated or controlled by the pump or motor of a system; but, in many systems, it is controlled by the variable restriction of a flow valve. So, before we get further into our study of hardware, we should look at the reaction of flow to restriction.

Restrictions used in flow control devices are like resistors in an electric circuit. Flow through a restriction increases as the pressure drop across it rises or as the restriction is reduced (the orifice size increased), while the pressure is held constant. Bilingual students—those who speak "electrical" as well as "hydraulics"—will recognize this as Ohm's law.

Parallel Restrictions

Figure 6.1a shows three restrictions, of various sizes, connected in parallel across a pumping system. Since the three restrictions (restricted passages) are connected in parallel, and only one pressure is applied, the pressure drop across each restriction is the same.

FIGURE 6.1
(a) Parallel restrictions.

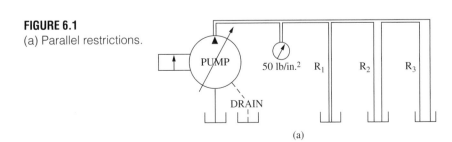

(a)

But, as stated earlier, the flow through the smaller passage (higher restriction) is less than that of the larger passages (lower restriction).

Note: Some may confuse parallel restrictions with check valves or relief valves connected in parallel. This is a different ball game from restrictions! To confuse check or relief valves with restrictions is like an Ohm's law student using Zener diodes and calling them resistors.

Check valves and relief valves connected in series and parallel may have their application, but in a different time and place. They have nothing to do with the restrictions used to measure and control volume of flow.

Some may think that, with parallel restrictions, all the flow will take the path of least restriction. But this is not so! The flow will divide, with the smallest flow going through the smallest passage. *Even when a very small diameter tube is paralleled with the largest sized pipe, some flow will still flow through the small tube.*

Figure 6.1b shows three resistors of different values connected in parallel across a battery. Because the resistors are in parallel, all voltages are the same. But the current through each resistor is inversely proportional to its resistance. This parallel electric circuit behaves the same as the hydraulic circuit of Figure 6.1a.

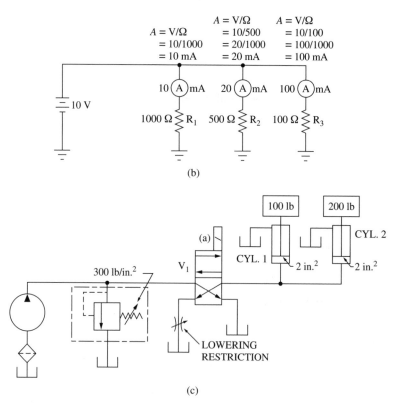

(b)

(c)

FIGURE 6.1 *(continued)* (b) Parallel resistors. (c) Parallel cylinders.

Parallel Cylinders

Figure 6.1c shows two identical cylinders connected in parallel. They are connected to a fixed-flow pump through a two-position valve. Cylinder 1 must lift a 100-lb load, and cylinder 2 is required to lift 200 lb.

Both cylinder caps are returned to tank through a variable lowering restriction when solenoid A is deenergized. Both rod ends are connected to tank. When solenoid A is energized, the upper window of V_1 moves down, connecting pressure from a fixed-flow pump to the cap ends of both cylinders. Fluid will try to move into both cylinder caps, but no flow can take place through either cylinder until enough pressure is built up to lift the load of that cylinder.

So, which cylinder will move first? Obviously it will be the one that requires the least pressure to move.

Cylinder 1 needs 100 lb/2 in.^{2} = **50 lb/in.^{2}**, whereas
Cylinder 2 needs 200 lb/2 in.^{2} = **100 lb/in.^{2}**

Cylinder 1 will lift its load as soon as the pressure reaches 50 lb/in.^{2}, but cylinder 2 cannot move until cylinder 1 is fully extended. The pressure will now build to 100 lb/in.^{2} allowing cylinder 2 to extend.

Some are confused by the action of Figure 6.1c. They think that all the flow takes the path of least restriction. It was not the restriction that caused cylinder 1 to extend first. *Cylinder 1 simply requires less pressure to lift its load.*

There are three levels of pressure in the operation of Figure 6.1c. Cylinder 1 moved during the first pressure level (50 lb/in.^{2}), cylinder 2 moved during the second level (100 lb/in.^{2}), and the pressure reached a third level of 300 lb/in.^{2} (at the relief valve setting) when cylinder 2 was fully extended.

Series Restrictions

Figure 6.2a shows three restrictions, of various sizes, connected in series across a pump. The flow is the same through all three restrictions, and the total restriction is equal to the sum of all three. The pressure drop across each restriction is proportional to its restriction, and the sum of all three pressure drops is equal to the pump pressure ($P_t = \Delta P_1 + \Delta P_2 + \Delta P_3$, where ΔP represents pressure drop).

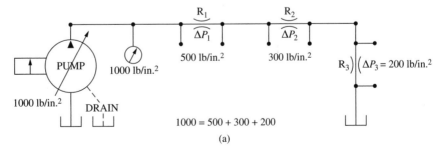

FIGURE 6.2
(a) Series restrictions.

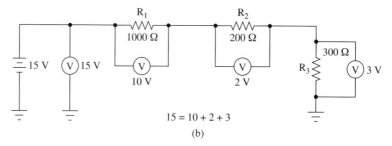

FIGURE 6.2 *(continued)*
(b) Series resistors.

Figure 6.2b shows three resistors of different values connected in series across a battery. The current is constant through all three resistors; and the total resistance is equal to the sum of all three resistances ($R_t = R_1 + R_2 + R_3$). The voltage drop across each resistor is directly proportional to its resistance; and the sum of all three voltage drops is equal to the battery voltage ($E_t = E_1 + E_2 + E_3$).

6.2   **NONCOMPENSATED FLOW CONTROL**

Figure 6.3 shows a cutaway view of a simple noncompensated flow control valve. The adjusting screw pushes a spool piston against a spring. At a certain position of the screw, the spool land opens to allow a restricted flow from port A to port B. Further opening of the spool land will allow more flow.

The spool of the valve also acts as a reverse-flow check valve. Notice that the metering land of the spool has a larger diameter than the sealing land. Because of the larger area of the metering land, pressure applied to port B will push the spool to the right, compressing the spring. When the spring is fully compressed, reverse flow will pass virtually unrestricted from port B to A.

While the forward flow through the valve of Figure 6.3 is proportional to a function of the adjusting screw, it is not generally referred to as a proportional valve since it is not electrically controlled.

Remote-Controlled Proportional Flow Control Valve

Figure 6.4 shows a cutaway drawing of a remotely controlled or proportional flow control valve. This valve is similar to that of Figure 6.3, except that it has an electrically operated force motor to push against one end of the spool while a spring pushes against the other.

The force motor is designed to produce force proportional to electric current (lb/A). Because the spring has a linear rate (in./lb), the opening of the spool will be directly proportional to current:

$$\frac{\text{in.}}{\cancel{\text{lb}}} \times \frac{\cancel{\text{lb}}}{\text{A}} = \text{in./A (inches of valve movement per ampere of electric current)}$$

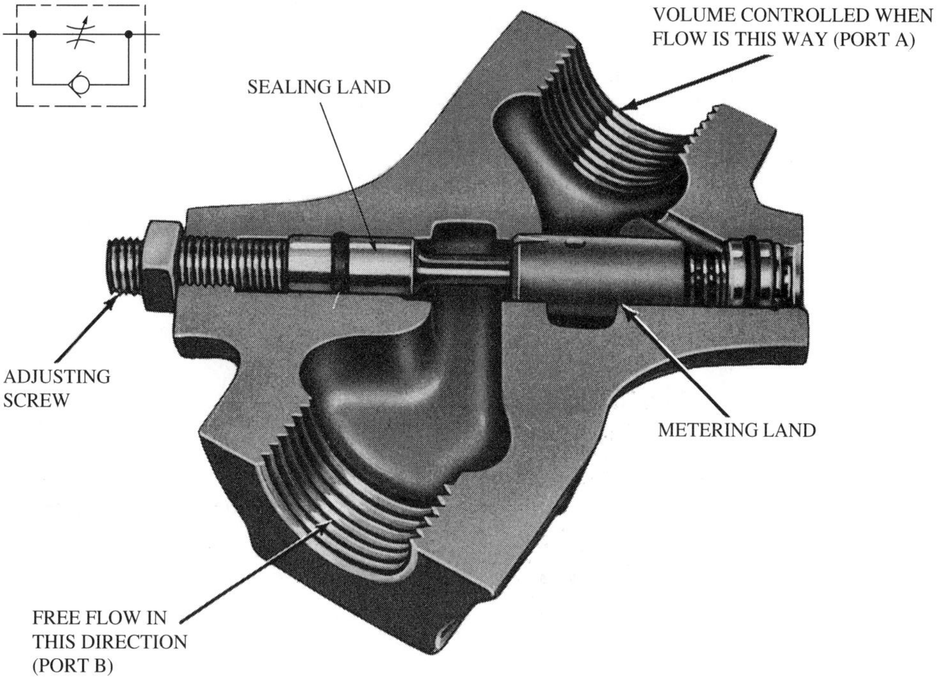

VOLUME CONTROLLED WHEN
FLOW IS THIS WAY (PORT A)

SEALING LAND

ADJUSTING
SCREW

METERING LAND

FREE FLOW IN
THIS DIRECTION
(PORT B)

FIGURE 6.3
Noncompensated flow control valve. (Courtesy of Vickers Inc.)

FIGURE 6.4
Basic remote-controlled
proportional valve. (Courtesy of
Vickers Inc.)

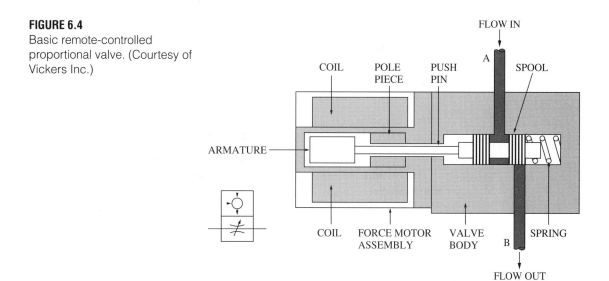

FLOW IN

COIL   POLE   PUSH   A   SPOOL
           PIECE   PIN

ARMATURE

COIL   FORCE MOTOR   VALVE   B   SPRING
            ASSEMBLY   BODY

FLOW OUT

6.3   THREE METHODS OF FLOW CONTROL

Figure 6.5 shows three methods of flow control: (a) meter-in, (b) meter-out, and (c) bleed-off. The choice of circuits depends on the application.

Meter-in Flow Control

Figure 6.5a shows a meter-in flow control. The pressure gauge P_S will register the relief valve or pump compensator setting. Gauge P_L will note the pressure needed to move the load when P_T is returned to tank at near-zero pressure.

The meter-in flow control circuit can only be used where there is no overrunning load or where a counterbalance valve has been installed to prevent overrun. Otherwise, the load would tend to run away.

To ensure power efficiency, the meter-in circuit is recommended for the system using a compensated pump. The compensated pump will stroke back to limit the flow

FIGURE 6.5
Three methods of flow control:
(a) Meter-in. (b) Meter-out.
(c) Bleed-off.

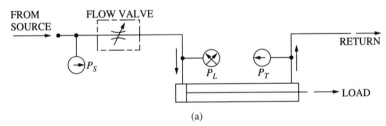

(a)

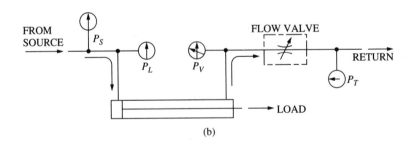

(b)

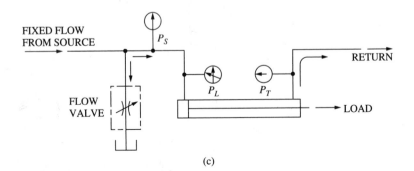

(c)

instead of dumping the excess flow to tank, as with a fixed-flow pump and relief valve. The meter-in flow control has good speed regulation.

Meter-out Flow Control

Figure 6.5b shows a meter-out flow control. Pressure gauges P_S and P_L will both register the supply pressure. The load pressure will be P_L minus P_v; and P_v will show the valve pressure when P_T is returned to tank at near-zero pressure.

The meter-out flow control will act as a counterbalance and is recommended where there is an overrunning load. The meter-out flow control is also recommended for a compensated pump system, for the same reasons as stated earlier for the meter-in flow control. The meter-out control has good speed regulation—perhaps a little better than the meter-in control, since it does not see any of the external cylinder leakage, but only the flow that moves the cylinder.

Bleed-off Flow Control

Figure 6.5c shows a bleed-off flow control. This circuit is used only with a fixed-flow pump. The control meters flow directly to tank, bypassing the load.

Assume the pump is delivering a fixed 10 gal/min and we want only 2 gal/min to move the load. We merely adjust the flow control to bleed off 8 gal/min leaving 2 gal/min for the load.

Pressure gauges P_S and P_L both register the load pressure when P_T is returned to tank at near-zero pressure.

The bleed-off flow control is the most power efficient of the three controls when used with a fixed-flow pump. The unused flow of the pump is returned to tank at the pressure of the load and not at the pressure of the relief valve.

The bleed-off flow control has the worst speed regulation of the three controls. With this method we control the part of the flow that is thrown away, leaving all the variables, such as leakages and pump speed changes, to affect the load speed. For this reason, the bleed-off should be limited to about 50% of pump flow.

6.4   EXTENDING VERSUS RETRACTING SPEEDS OF CYLINDERS

Cylinders may be used in systems with either fixed-flow or fixed-pressure supplies. Section 3.1 of Chapter 3 (Figure 3.2) shows the extending versus retracting speeds of a differential cylinder when supplied by a fixed-flow source. Before we tackle cylinder systems supplied by fixed-pressure sources, let's have another look at cylinder systems using fixed-flow sources.

■   ### EXAMPLE 6.1: 2:1 Cylinders with Fixed-Flow Sources (U.S. Units)

Figure 6.6 shows a retracting and extending cylinder system supplied with a fixed-flow source. When solenoid A of Figure 6.6a is deenergized, the full 800 in.^{3}/min delivery of the pump is ported to the cap end of the cylinder, which has an area of 4 in.^{2}. The

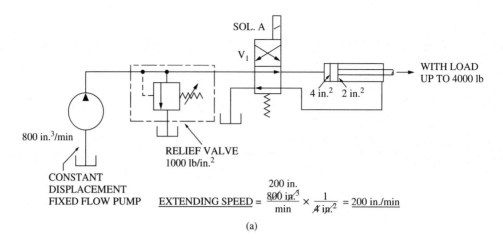

$$\text{EXTENDING SPEED} = \frac{\overset{200\ in.}{\cancel{800\ in.^3}}}{min} \times \frac{1}{\cancel{4}\ in.^2} = \underline{200\ in./min}$$

(a)

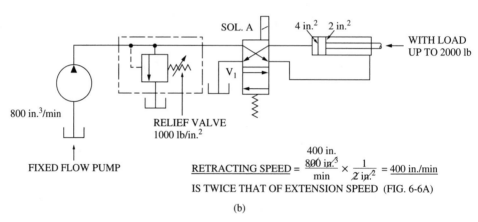

$$\text{RETRACTING SPEED} = \frac{\overset{400\ in.}{\cancel{800\ in.^3}}}{min} \times \frac{1}{\cancel{2}\ in.^2} = \underline{400\ in./min}$$

IS TWICE THAT OF EXTENSION SPEED (FIG. 6-6A)

(b)

FIGURE 6.6
Cylinder (a) extension and (b) retraction with fixed-flow pump.

cylinder will extend. The flow from the rod end of the cylinder is ported back through V_1 to tank.

Solution: To calculate the extending speed of the cylinder, merely arrange the dimensions (along with their numerical values) until all but "in./min" are cancelled. The formula for those who need it is Speed = Flow × 1/Area:

$$\text{Extension speed} = \frac{\overset{200\ in.}{\cancel{800\ in.^3}}}{min} \times \frac{1}{\cancel{4}\ in.^2} = \mathbf{200\ in./min}$$

When solenoid A of Figure 6.6b is energized, the 800 in.³/min pump delivery will be directed to the active (annulus) area of the cylinder (2 in.²) (calculated by subtracting the area of the rod from the area of the piston). The cylinder will retract, with the flow from the cap end returning to tank through V_1. The retracting speed can also be calculated by arranging the dimensions so that all are eliminated except "in./min." (Speed = Flow × 1/Area.)

$$\text{Retraction speed} = \frac{\overset{400 \text{ in.}}{\cancel{800 \text{ in.}^3}}}{\min} \times \frac{1}{\cancel{2 \text{ in.}^2}} = \textbf{400 in./min} \qquad \blacksquare$$

■   **EXAMPLE 6.2: 2:1 Cylinders with Fixed-Flow Sources (SI Units)**

Consider the extension versus retraction of a 2:1 cylinder using the following SI units for Figure 6.6:

1. Cylinder cap-end area = 40 cm^2.
2. Cylinder rod-end annulus area = 20 cm^2.
3. Pump flow = 8 L/min = 8000 cm^3/min.

Solution

$$\text{Extension speed} = \frac{\overset{200 \text{ cm}}{\cancel{8000 \text{ cm}^3}}}{\min} \times \frac{1}{\cancel{40 \text{ cm}^2}} = \textbf{200 cm/min}$$

$$\text{Retraction speed} = \frac{\overset{400 \text{ cm}}{\cancel{8000 \text{ cm}^3}}}{M} \times \frac{1}{\cancel{20 \text{ cm}^2}} = \textbf{400 cm/min}$$

So, with a fixed flow from the pump, the single-rod differential cylinder with a 2:1 area ratio will retract twice as fast as it extends.   ■

Cylinders with Fixed-Pressure Sources

Figure 6.7 shows an extension and retraction system operating a nonloaded cylinder. The pressure is fixed by a pressure-compensated pump. A noncompensated flow control valve (V_1) is connected in a meter-in arrangement between the pump and the directional valve (V_2). The pump has sufficient flow, and the flow valve (V_1) sufficient restriction, to allow the pump to maintain its fixed pressure of 1000 lb/in.^{2} at all times.

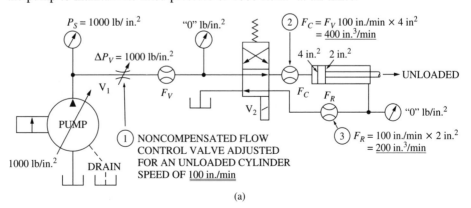

(a)

FIGURE 6.7
Fixed-pressure, meter-in, noncompensated flow control of an unloaded cylinder.
(a) Extension.

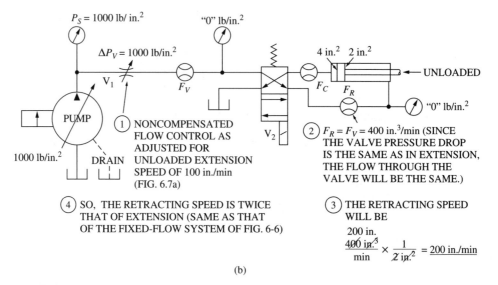

(b)

FIGURE 6.7 *(continued)*
Fixed-pressure, meter-in, noncompensated flow control of an unloaded cylinder.
(b) Retraction.

When the solenoid of V_2 is energized, as shown by Figure 6.7a, the pump is applied through the flow valve V_1, flow meter F_v, directional valve V_2, and flow meter F_c to the cap end of the unloaded cylinder. The rod end of the cylinder is returned to tank through the directional valve. The flow valve is adjusted for a cylinder speed of 100 in./min. The cap-end flow (F_h) calculates to:

$$F_c = \frac{100 \text{ in.}}{\text{min}} \times 4 \text{ in.}^2$$

$$= 400 \text{ in.}^3/\text{min}$$

The rod-end flow figures to:

$$F_r = \frac{100 \text{ in.}}{\text{min}} \times 2 \text{ in.}^2$$

$$= 200 \text{ in.}^3/\text{min}$$

Since the cylinder is unloaded, we assume that no energy is taken by the cylinder. All energy taken from the pump is used to push the 400 in.³/min through the flow valve at a pressure drop of 1000 lb/in.².

When the solenoid of V_2 is deenergized, as shown by Figure 6.7b, the pump is applied through the flow valve V_1, flow meter F_v, directional valve V_2, and flow meter F_r to the rod end of the unloaded cylinder. The cap end of the cylinder is returned to tank through the directional valve.

As before, all energy is assumed to be used to push 400 in.³/min through the flow valve at a pressure drop of 1000 lb/in.². Since the pressure drop across the flow valve during retraction is the same as it was for extension, the flow through it must also be the same (400 in.³/min). Since the 400 in.³/min is now flowing into the rod end of the cylinder, however, the retracting speed is:

$$\text{Retraction speed} = \frac{\overset{200 \text{ in.}}{\cancel{400 \text{ in.}^3}}}{\text{min}} \times \frac{1}{\cancel{2 \text{ in.}^2}} = \textbf{200 in./min}$$

(The retracting speed of 200 in./min is twice the extending rate set by the flow valve.) *So, with a meter-in noncompensated flow valve and a fixed-pressure system, the unloaded 2:1 cylinder will retract twice as fast as it extends* (as was true for the fixed-flow system of Figure 6.6).

Figure 6.8 shows a fixed-pressure system driving a meter-out, noncompensated flow control of an unloaded cylinder. When the solenoid of V_2 is energized, as shown in Figure 6.8a, the pump pressure of 1000 lb/in.² is applied through the directional valve V_2 directly to the cap-end area of the unloaded cylinder. This action accelerates the cylinder until it reaches the speed set by the flow valve. At this time, a strange thing happens. *The rod-end pressure (P_r) rises to 2000 lb/in.², supplying twice the pressure to the flow valve as that supplied by the pump.* The unloaded cylinder has transformed, or *intensified,* the pressure from 1000 to 2000 lb/in.².

Now, there are at least three related physical laws that explain what happens here as well as, at least, one valid electrical analogy (shown later): (1) *Power in* equals *power out,* (2) *Energy in* equals *energy out,* and (3) Newton's third law of motion which, when paraphrased to fit a hydraulics cylinder, states "When a cylinder is in motion but is not accelerating or decelerating, the extending force (lbs) must equal the retracting force." Since the cylinder of Figure 6.8a is not accelerating and there is no external load, the extending hydraulic force must equal the retracting hydraulic force.

The extending hydraulic force on the cylinder is:

$$\text{Cap-end force} = \frac{1000 \text{ lb}}{\cancel{\text{in.}^2}} \times 4 \, \cancel{\text{in.}^2} = \textbf{4000 lb}$$

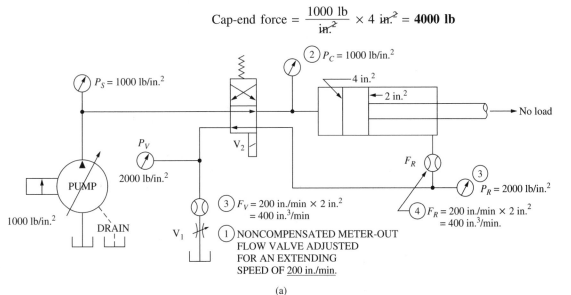

(a)

FIGURE 6.8
Fixed-pressure, meter-out, noncompensated flow control of an unloaded cylinder. (a) Extension.

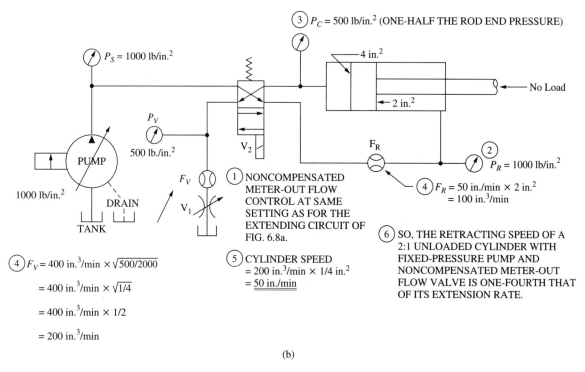

(b)

FIGURE 6.8 *(continued)*
Fixed-pressure, meter-out, noncompensated flow control of an unloaded cylinder. (b) Retraction.

and the pressure on the rod end must be:

$$\text{Rod-end pressure} = \frac{\dfrac{2000}{4000 \text{ lb}}}{2 \text{ in.}^2} = \mathbf{2000 \text{ lb/in.}^2}$$

An extending, unloaded, single-rod, constant-velocity cylinder with a 2:1 area ratio will deliver twice the pressure from its rod end (2000 lb/in.^{2}) as is supplied to its head end (1000 lb/in.^{2}). If the area ratio had been 4:1, the rod-end pressure would have been 4000 lb/in.^{2}. Such a cylinder is called an intensifier.

Let's leave the analysis of Figure 6.8 at this point while we look at the electrical analogy of a hydraulic intensifier.

Electrical Analogy of Hydraulic Intensifier and Deintensifier

Figure 6.9a shows a step-up transformer, with a 2:1 turns ratio, connected between an alternator and a 55-Ω resistor. With 110 V$_{ac}$ applied to the 110 turns of the transformer primary, the 220-turn secondary will deliver 220 volts.

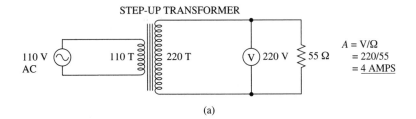

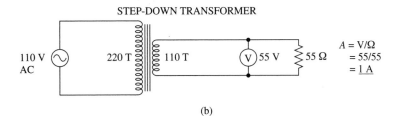

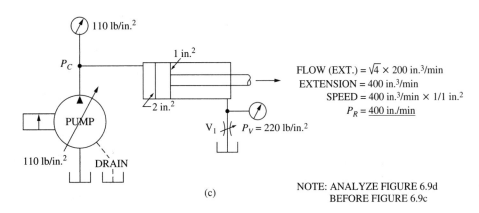

NOTE: ANALYZE FIGURE 6.9d
BEFORE FIGURE 6.9c

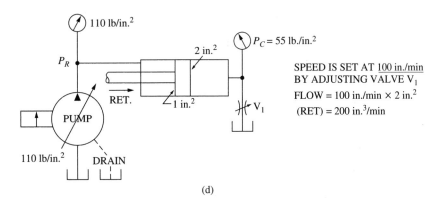

FIGURE 6.9
Electrical analogy of (a) step-up and (b) step-down hydraulic intensifier and deintensifier.
(c) Equivalent hydraulic intensifier (step-up) and (d) deintensifier (step-down).

Note: A voltage of 110 V_{ac} is applied to the 110 turns of the transformer primary. This is 1 V/turn for the primary and for all other turns sharing the same magnetic field. The secondary has 220 turns sharing the same magnetic field as the primary; therefore, the secondary has 220 V.

Applying Ohm's law, the current through the 55-Ω resistor, using the step-up transformer, is A = V/Ω = 220/55 = **4 amps.**

Figure 6.9b shows the same transformer turned around to step down the voltage applied to the resistor. Now, since each turn of the primary—and hence all turns of the transformer—will see only 0.5 V per turn and the secondary will apply only 55 V to the resistor, then

$$\text{Secondary voltage} = 110\ \cancel{T} \times \frac{0.5\ \text{V}}{\cancel{T}} = \textbf{55 V}$$

Again, applying Ohm's law, the current through the resistor, using the step-down transformer, is A = V/Ω = 55/55 = **1 amp.**

Figure 6.9c shows an intensifier (step-up) cylinder and Figure 6.9d a deintensifier (step-down) cylinder. As long as the area ratio of the cylinder is the same as the turns ratio of the transformer (2:1), the hydraulic circuits of Figures 6.9c and 6.9d function the same as the electric circuits of Figures 6.9a and 6.9b.

In Figure 6.9c the pressure source P_s (110 lb/in.^{2}) is intensified to 220 lb/in.^{2} and appears across the flow valve V_1 during extension. The 110 lb/in.^{2} source is deintensified to 55 lb/in.^{2} and appears across the same flow valve during retraction (see Figure 6.9d).

Laminar Flow

If the flow through the valve is laminar (i.e., there are no sharp-edged orifices inside the valve to cause turbulence), *it will be directly proportional to the pressure drop across it.*

Let's assume that the flow is laminar and the valve is adjusted for a cylinder speed of 100 in./min during retraction (Figure 6.9d). The flow through the valve will be:

$$\text{Retraction flow} = \frac{100\ \text{in.}}{\text{min}} \times 2\ \text{in.}^2 = \textbf{200 in.}^3\textbf{/min}$$

Let's turn the cylinder around, as shown in Figure 6.9c without disturbing the valve setting. How fast will the cylinder extend? Given

1. The flow through the laminar valve is directly proportional to its pressure drop.
2. The flow through the valve was 200 in.^{3}/min with 55 lb/in.^{2} pressure drop.
3. The pressure drop across the valve during cylinder extension is 220 lb/in.^{2}.

Solution: Since the pressure across the valve during cylinder extension (220 lb/in.^{2}) is four times what it was for cylinder retraction (55 lb/in.^{2}), the flow will also be four times as much. So,

$$\text{Extension flow} = 4 \times 200\ \text{in.}^3\text{/min} = \textbf{800 in.}^3\textbf{/min}$$

and the extending speed of the cylinder will be:

$$\text{Extension speed} = \frac{800 \; \cancel{\text{in.}}^3}{\min} \times \frac{1}{1 \; \cancel{\text{in.}}^2} = \textbf{800 in./min}$$

So, with a fixed supply pressure, a noncompensated, laminar-type, meter-out flow control will cause an unloaded 2:1 area cylinder to extend eight times as fast as it retracts. With laminar flow:

Retraction speed = **100 in./min**

Extension speed = **800 in./min**

Yes, we said "It extends faster than it retracts," in contrast to the case of the fixed-flow system where it retracts faster than it extends.

Turbulent Flow

Now, in practice, it is very difficult to design a control valve without sharp-edged orifices. Therefore, most valves create turbulence, which destroys the linear relationship between the flow and pressure drop. *In a turbulent flow valve, the flow is approximately proportional to the square root of the pressure drop across the restriction.*

If we assume that the valve of Figures 6.9c and 6.9d has turbulent flow, we must apply the square-root-of-pressure law to the flows. Since the pressure drop across the valve during extension (220 lb/in.^{2}) is four times that of retraction (55 lb/in.^{2}), the flow through the valve during extension will be: $\sqrt{4} = 2$ times 200 in.^{3}/min = **400 in.^{3}/min,** and the speed of extension will be:

$$\text{Extension speed} = \frac{400 \; \cancel{\text{in.}}^3}{\min} \times \frac{1}{1 \; \cancel{\text{in.}}^2} = \textbf{400 in./min}$$

So, with a fixed supply pressure, a noncompensated, *turbulent-type,* meter-out flow control will cause an unloaded 2:1 area cylinder to extend four times as fast as it retracts. With turbulent flow:

Retraction speed = 100 in./min

Extension speed = 400 in./min

Now, let's go back to the system of Figure 6.8 to see what happens to the relative speeds of this unloaded cylinder. Remember that it had a fixed pump pressure source of 1000 lb/in.^{2} and a noncompensated flow control valve connected in a meter-out circuit. Since the valve is a practical design, with sharp-edged orifices, we must assume turbulent flow.

After the cylinder of Figure 6.8a has stopped accelerating and there is no external load, the rod end, with half the area (2 in.^{2}), must see twice the pressure (P_r = 2000 lb/in.^{2}). This places 2000 lb/in.^{2} pressure across the flow valve (V_1), which is adjusted to provide an extending speed of 200 in./min. This sets the flow through V_1 at:

$$\frac{200 \; \text{in.}}{\min} \times 2 \; \text{in.}^2 = \textbf{400 in.}^3\textbf{/min}$$

Figure 6.8b shows the retracting portion of the cylinder cycle when V_2 is deenergized. The flow valve has the same setting, which provided 200 in./min during extension. The flow through the flow valve is now:

$$F_v = 400 \text{ in.}^3/\text{min (flow through valve during extension)} \times \sqrt{(500/2000)}$$
$$= 400 \text{ in.}^3/\text{min} \times \sqrt{(1/4)}$$
$$= 400 \text{ in.}^3/\text{min} \times 1/2$$
$$= \mathbf{200 \text{ in.}^3/\text{min}}$$

and the speed during retraction is:

$$\text{Retraction speed} = \frac{50 \text{ in.}}{\dfrac{200 \text{ in.}^3}{\text{min}}} \times \frac{1}{4 \text{ in.}^2} = \mathbf{50 \text{ in./min}}$$

So, the retracting speed of a 2:1 unloaded cylinder, with a fixed-pressure pump and a meter-out noncompensated flow valve, is one-fourth that of its extension rate.

Cylinder Circuits with Extension Speeds of 1.414 Times Their Retraction Rates

Figure 6.10 shows a cylinder cycling circuit where the noncompensated flow valve V_1 is located on the load side of the directional valve V_2, in series with the rod end. *This flow valve must be symmetrical to control flow equally in both directions.*

The flow valve is adjusted for a retracting speed of 100 in./min, as shown in Figure 6.10a. The fluid from the cap end of the cylinder is returned to tank through V_2. This causes both the cap-end and the rod-end pressures to stay at zero, with the flow valve seeing the full 1000 lb/in.^{2} during retraction.

Note: The cylinder can be neither an intensifier nor a deintensifier if either end is connected to tank. In the electrical analogy of an intensifier, the transformer would not produce voltage either if it were shorted out.

Due to the 2:1 pressure intensification during extension of the cylinder in Figure 6.10b the pressure across the flow valve (V_1) will be 2000 lb/in.^{2}.

Since the pressure across the flow valve has doubled, the extending speed of the cylinder will be $\sqrt{2} \times 100$ in./min = $1.414 \times 100 = \mathbf{141.4 \text{ in./min.}}$

So, with a fixed pressure 2:1 cylinder cycling system, and a noncompensated flow valve connected between the rod end of the cylinder and the reversing valve, the extending speed will be 1.414 times the retracting speed.

If the flow valve of Figure 6.10 were removed and placed between the cap end of the cylinder and the reversing valve V_2, the cylinder would still extend 1.414 times its retracting rate. (Try it.)

In this case the cylinder would act as a deintensifier during retraction, providing only half the pressure to the valve. With half pressure during retraction, the cylinder would still extend 1.414 times its retracting rate.

Many servo and other proportional valves have enough restriction, and the pump enough flow capacity, to maintain a fixed pressure setting, even with the valve wide open.

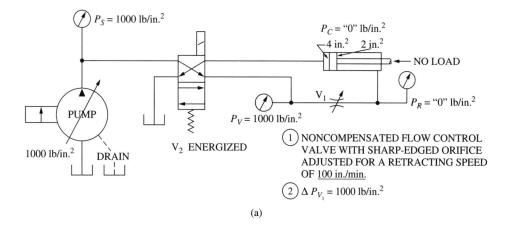

(a)

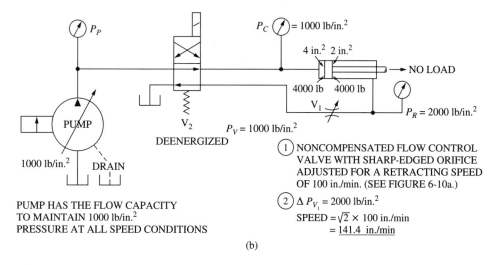

PUMP HAS THE FLOW CAPACITY
TO MAINTAIN 1000 lb/in.^{2}
PRESSURE AT ALL SPEED CONDITIONS

(b)

FIGURE 6.10

Flow valve on load side of directional valve. (a) Retraction. (b) Extension.

Figure 6.11 shows a modified drawing of a proportional valve connected between a pressure-compensated pump and cylinder. This servo or proportional valve has four-way operation as well as variable restrictions.

Note: See Chapter 9 for more on servo valves.

Figure 6.11a shows the retracting cylinder and Figure 6.11b depicts its extension. It is assumed that the valve is electrically saturated (wide open) in each direction, providing maximum speeds.

In the analysis of Figure 6.11, dimensional numbers are selected that satisfy all conditions of retraction and extension. The overall system requirements are (1) a fixed

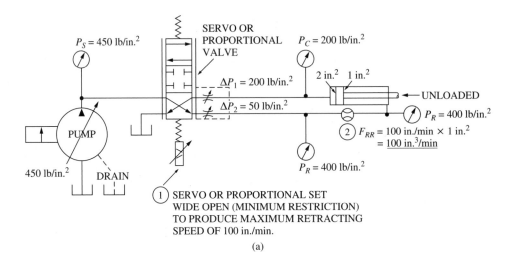

(a)

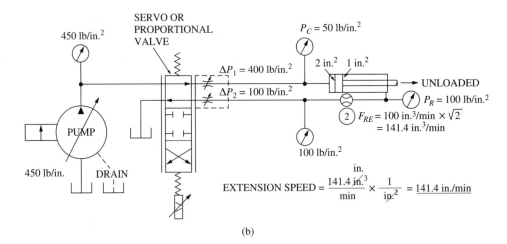

(b)

FIGURE 6.11
Maximum (a) retraction and (b) extension rates of servo and proportional valves.

supply pressure of 450 lb/in.^{2} and (2) sufficient pump flow capacity to maintain the 450 lb/in.^{2}.

Additional information during retraction (Figure 6.11a):

1. ΔP_2 (50 lb/in.^{2}) plus P_r (400 lb/in.^{2}) = P_s (450 lb/in.^{2}). *Note:* Pressures "upstream" of the cylinder (from the pump to the cylinder) must add up to the supply pressure. Those pressures beyond ("downstream") of the cylinder are generated by the deintensifier action of the cylinder.
2. P_c = 200 lb/in.^{2} (one-half P_r). This is due to the deintensifier action of the 2:1 unloaded cylinder.

3. $\Delta P_1 = P_c = 200$ lb/in.^{2} (obvious).
4. Cylinder speed = 100 in./min (assumed).
5. Retracting rod-end flow (F_{rr}) = 100 in./min × 1 in.^{2} = 100 in.^{3}/min.

Additional information during extension (Figure 6.11b):

1. ΔP_1 (400 lb/in.^{2}) plus P_c (50 lb/in.^{2}) = P_s (450 lb/in.^{2}).
2. $P_r = 100$ lb/in.^{2} (twice P_c). This is due to the intensifier action of the 2:1 unloaded cylinder.
3. $\Delta P_2 = P_r = 100$ lb/in.^{2} (obvious).
4. Observe that the pressure drops across the valve are double what they were during retraction. *During retraction:* $\Delta P_1 = 200$ lb/in.^{2} and $\Delta P_2 = 50$ lb/in.^{2}. *During extension:* $\Delta P_1 = 400$ lb/in.^{2} and $\Delta P_2 = 100$ lb/in.^{2}.
5. Since the pressures across the valve were doubled during extension, the flows and hence the speed will be increased by the square root of 2 or 1.414.
6. The extending speed will be 100 in./min × 1.414 = **141.4 in./min.**

So, when a servo valve or a proportional valve is cycled at saturated speed, the extension speed of a 2:1 cylinder will be 1.414 times the retracting rate.

Ohm's Law for Hydraulics?

Ohm's law was derived from a basic physical principle that states: "The result equals the effort divided by the opposition." Or restated, "The result is directly proportional to the effort and inversely proportional to the opposition." These "statements of fact" are obvious and apply to hydraulics as well as to electronics.

We could have an Ohm's law for hydraulics, as well as electronics, if units were devised where a unit of flow (say, "F") were equal to some unit of pressure (say, "P") divided by some unit of restriction (say, "R"), or $F = P/R$.

Even though electrical technologists have always used hydraulic analogies to better understand electricity, the hydraulics industry has failed, thus far, to adopt an Ohm's law for hydraulics. Some have tried and have introduced the unit "Lohm" which is the rate of flow per unit of pressure, just as the ohm is amps per volt in electronics.

Because the principles of hydraulics and electronics are so similar, any knowledge you may have of Ohm's law or other electrical principles will enhance your knowledge of hydraulics.

6.5   PRESSURE-COMPENSATED FLOW CONTROLS

If a flow control is to regulate the speed of a motor or cylinder properly, it must hold the flow constant at any flow setting. It must do this regardless of load changes and cylinder intensification or deintensification.

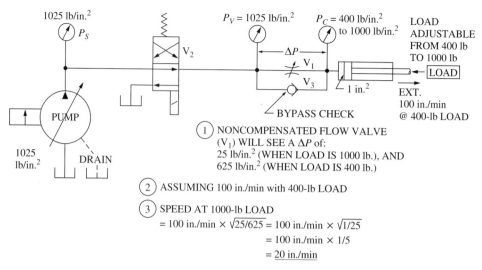

FIGURE 6.12
Noncompensated flow control with varying load.

Need for Pressure Compensation

Figure 6.12 shows a cylinder system with a noncompensated flow valve trying to control the speed of a varying load. The load has an opposing force that varies from 400 to 1000 lb during the forward travel of the cylinder.

When the cylinder is extending against a load force of 400 lb, the 1-in.^{2} cap end of the cylinder sees 400 lb/in.^{2}. When the opposing load changes to 1000 lb, the cap-end pressure rises to 1000 lb/in.^{2}.

Since the pump pressure is fixed at 1025 lb/in.^{2}, the pressure drop across the flow valve is as follows:

$$1025 - 1000 = \textbf{25 lb/in.}^2 \text{ across valve, when the opposing load is 1000 lb}$$

$$1025 - 400 = \textbf{625 lb/in.}^2 \text{ across valve, when the opposing load is 400 lb}$$

Remember that with a noncompensated flow control valve using sharp-edged orifices, the flow is proportional to the square root of the pressure drop across the valve. The ratio of pressure drops is 625/25 = 25:1. The square root of 25 is 5. So, with the 400-lb load, the cylinder speed will be five times the speed with the 1000-lb load. If the valve is adjusted to provide an extending cylinder speed of 100 in./min with the 400-lb load, the cylinder will slow down to 100/5 = 20 in./min when it encounters the 1000-lb load. This is poor speed regulation.

Flow Control with Built-in Magic Box

A good way to keep the flow from changing, when the cylinder or motor sees a change in load pressure, is to keep the pressure drop across the valve throttle constant. For simplification purposes, let's first use a "magic box" to do this. We will later see what's in the magic box.

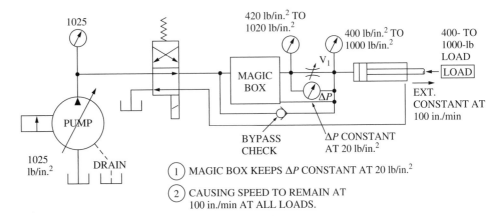

FIGURE 6.13
Varying load with magic-box control.

In the example of Figure 6.13, the magic box senses the pressure drop across the variable orifice (throttle) (V_2), and keeps it constant at 20 lb/in.^{2}. As the pressure at the cap end of the cylinder (P_c) changes from 400 lb/in.^{2} to 1000 lb/in.^{2} due to a load increase, the pressure at the input to the throttle (P_v) changes from 420 lb/in.^{2} to 1020 lb/in.^{2}, keeping the pressure across the throttle fixed at 20 lb/in.^{2}.

Because the pressure across the throttle is fixed, the flow to the cylinder will change only as the throttle is changed, but will remain constant when the cylinder load increases from 400 to 1000 lb. If the valve throttle is set to provide a cylinder speed of 100 in./min with a load of 400 lb, it will remain at 100 in./min when the load is increased to 1000 lb.

Restrictive Pressure-Compensated Flow Control

Figure 6.14 shows us what's inside the magic box to keep the pressure across the throttle constant, even when the load pressure changes. The compensator spool does the trick. *When the load pressure changes, the compensator spool alters its own pressure drop to keep a constant pressure across the throttle.*

The fluid route through the flow control is as follows: It goes *in* through the compensator restriction, to the center of the spool, and *out* through the throttle to the load cylinder. The compensator spool sees the same pressure drop as the throttle.

The right end of the throttle is connected to the top end of the spool, while the left end of the throttle is connected to the bottom. A 20-lb spring, in addition to the pressure from the right end of the throttle, acts against the top end of the spool.

The action of the compensator is automatic. The end areas of the spool are equal. So the pressure on top of the spool, plus the equivalent spring pressure (20 lb/in.^{2}), must remain equal to the pressure at the bottom of the spool, or the spool will move until this condition is satisfied. The pressure across the spool, and across the throttle, will remain at 20 lb/in.^{2} (equal to the equivalent spring pressure).

When the pressure across the throttle of a flow control is maintained at a constant value, the flow will be directly related to the opening of the throttle and completely independent of the load pressure.

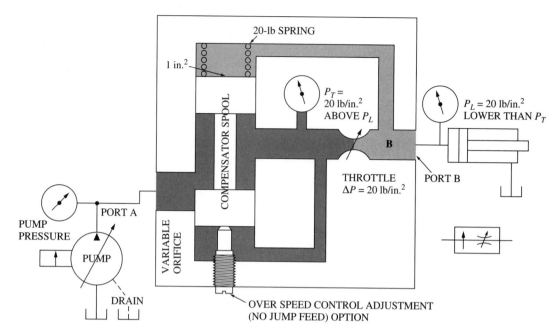

FIGURE 6.14
Restrictive pressure-compensated flow control. (Courtesy of Vickers Inc.)

The restrictive-type flow control is better suited for the system using a pressure-compensated pump. The pump compensator will limit the flow to the amount needed by the load and, thus, avoid wasted flow to tank. This will improve the overall efficiency of the system.

When the system of Figure 6.14 is turned off, the spring at the top of the compensator spool will push the spool down slightly until it rests against an adjustable plunger called an *overspeed control*. Without this stop, the spool would have been pushed all the way to the bottom, reducing the restriction to the throttle, when the system is first restarted. This could cause the load to lurch forward until the flow valve is under control.

Bypass-type Pressure-Compensated Flow Control

Figure 6.15 shows a bypass-type pressure-compensated flow control. Instead of a balanced restrictive spool, this control has a balanced plunger piston, which meters flow to tank in order to maintain a constant pressure across the throttle. This action is much like that of a relief valve.

This spring, plus pressure, on the lower end of the plunger is equal to the pressure at the top end. The pressure difference across the plunger, equal to the equivalent pressure of the spring (20 lb/in.^{2}), is also connected across the throttle. If anything should upset this balance, the plunger would open or close slightly to bring the piston back into balance and restore the pressure across the throttle to 20 lb/in.^{2}.

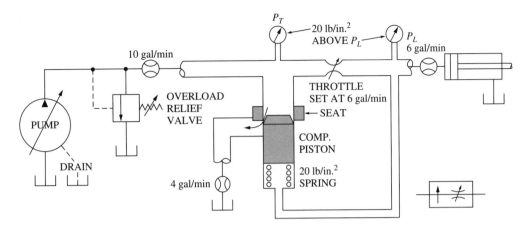

FIGURE 6.15
Bypass-type pressure-compensated flow control.

The bypass-type control limits the system pressure to that of the load, plus 20 lb/in.² for the throttle. For this reason it is best suited for use with a constant flow system. The overload relief valve is added to protect the system against excessive pressure when the cylinder bottoms out.

Some bypass-type designs use the compensator piston to double as an overload relief valve by adding a spring-loaded poppet from the bottom end of the piston to tank. With this arrangement the valve acts as a balanced-piston relief valve as well as pressure compensator.

Temperature-Compensated, Pressure-Compensated Flow Control

Figure 6.16 shows a temperature-compensated version of the restrictive-type pressure-compensated flow control. The control is the same as that of Figure 6.14 except that it has a special throttle design. A temperature-compensating rod connects the throttle control to a metering land shaped like a notch-rimmed cup.

As the temperature of the fluid goes up, its viscosity goes down, allowing more hot thin oil to pass through the same orifice. When this happens in the control of Figure 6.16, the temperature-compensating rod will expand, slightly closing the throttle, to bring the flow back to its intended rate. The notched metering land provides a more gradual change of restriction versus temperature.

The temperature-compensated throttle can also be used in the bypass-type flow control with the same results.

We now have the ultimate in flow control devices. *With the temperature-compensated pressure-compensated flow control, the speed of a hydraulically operated device will remain constant, regardless of the oil temperature, load change, or intensification or deintensification of the oil.*

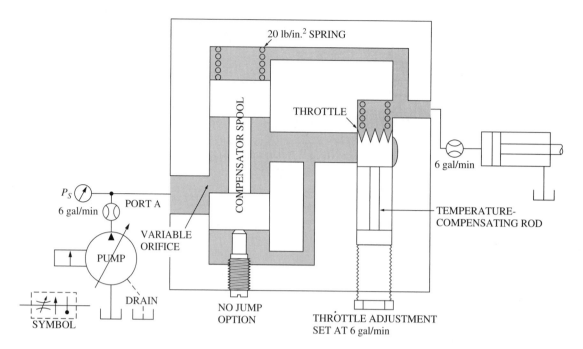

FIGURE 6.16

Temperature-compensated, pressure-compensated flow control. (Courtesy of Vickers Inc.)

6.6   SUMMARY

1. Restrictions used in flow control devices are like resistors in an electric circuit.
2. Flow through a restriction increases, either as the pressure drop across it rises or as the restriction is reduced.
3. When three restrictions, of different sizes, are connected in parallel, the pressure drop across each is the same, but the flow through the smaller passage (higher restriction) is less than that of the larger passages (lower restriction).
4. When three restrictions of different sizes are connected in series across a pump, the flow is the same through all three restrictions, and the total restriction is equal to the sum of all three.
5. The proportional remote-controlled valve is similar to the manually operated noncompensated flow control, except that the remote valve has an electrically operated force motor to push against one end of the spool while a spring pushes against the other.
6. To ensure power efficiency, the meter-in and meter-out circuits are recommended for systems using compensated pumps.
7. The meter-out flow control will act as a counterbalance and is recommended where there is an overrunning load.

8. The bleed-off flow control is the most power efficient of the three controls when used with a fixed-flow pump.

9. The bleed-off flow control has the worst speed regulation of the three controls.

10. With a fixed flow from the pump, the single-rod differential cylinder with a 2:1 area ratio will retract twice as fast as it extends.

11. Newton's third law of motion, when paraphrased to fit the hydraulics cylinder, states: "When a cylinder is in motion but is not accelerating or decelerating, the extending force must equal the retracting force."

12. An extending, unloaded, single-rod, constant-velocity cylinder with a 2:1 area ratio will deliver twice the pressure from its rod end as is supplied to its cap end.

13. The step-up transformer is the electrical equivalent of the hydraulic intensifier.

14. If the flow through the valve is laminar (no sharp-edged orifices inside the valve to cause turbulence), the flow through the valve will be directly proportional to the pressure drop across it.

15. With a fixed supply pressure, a noncompensated, *laminar-type,* meter-out flow control will cause an unloaded 2:1 area cylinder to extend *eight* times as fast as it retracts.

16. In a turbulent flow valve, the flow is approximately proportional to the square root of the pressure drop across the valve.

17. With a fixed supply pressure, a noncompensated, *turbulent-type,* meter-out flow control will cause an unloaded 2:1 area cylinder to extend *four* times as fast as it retracts.

18. With a fixed-pressure 2:1 cylinder cycling system, and a noncompensated flow valve connected between the rod end of the cylinder and the reversing valve, the extending speed will be 1.414 times the retracting speed.

19. When a servo valve or a proportional valve is cycled at saturated speed, the extending speed will be 1.414 times the retracting rate.

20. We could have an Ohm's law for hydraulics, as well as electronics, if units were devised where a unit of flow (say, "F") were equal to some unit of pressure (say, "P") divided by some unit of restriction (say, "R"), or $F = P/R$.

21. Because the principles of hydraulics and electronics are so similar, any knowledge you may have of Ohm's law or other electrical principles will enhance your knowledge of hydraulics.

22. If a flow control is to regulate the speed of a motor or cylinder properly, it must hold the flow constant at any flow setting, regardless of the load changes.

23. When the valve throttle of a pressure-compensated flow control is set to provide a cylinder speed of 100 in./min with a load of 400 lb, it will remain at 100 in./min when the load is increased to 1000 lb.

24. When the load pressure changes, the compensator spool alters its own pressure drop to keep a constant pressure across the throttle.

25. When the pressure across the throttle of a flow control is maintained at a constant value, the flow will be directly related to the opening of the throttle and completely independent of the load pressure.

26. With the temperature-compensated, pressure-compensated flow control, the speed of a hydraulically operated device will remain constant regardless of the oil temperature, load change, or intensification or deintensification of the oil.

6.7 PROBLEMS AND QUESTIONS

1. True or false?: In most systems, flow is controlled by the variable restriction of a valve.
2. True or false?: In a parallel circuit, all the flow will take the path of least restriction.
3. True or false?: In the electrical analogy of hydraulics, the restriction is like a Zener diode.
4. True or false?: When resistors are in parallel, all voltages are the same.
5. True or false?: When resistors are connected in series, all voltages are the same.
6. True or false?: To ensure power efficiency, the meter-in and meter-out circuits are recommended for systems using fixed-flow pumps.
7. True or false?: The bleed-off flow control is the most power efficient of the three flow controls when used with a fixed-pressure pump.
8. True or false?: The bleed-off control has the best speed regulation of the three controls.
9. True or false?: The manually operated flow valve is not referred to as a proportional valve since it is not electrically operated.
10. True or false?: The spool of the valve of Figure 6.3 also acts as a reverse-flow check valve.
11. True or false?: Cylinders can only be used in systems with fixed-flow power supplies.
12. From Figure 6.6a, what is the flow from the rod end of the cylinder while extending?
13. From Figure 6.6b, what is the flow from the cap end of the cylinder while retracting?
14. From Figure 6.7a, what is the flow through the flow valve (F_v) during extension?
15. From Figure 6.7b, what is the flow from the cap end of the cylinder (F_c) during retraction?
16. From Figure 6.8, recalculate the retracting speed if the extending speed is increased to 400 in./min. Show work.
17. From Figure 6.9b, what would be the flow through the 55-Ω resistor if the ac input to the primary of the transformer were increased to 120 V?
18. From Figure 6.9d what would be the cap-end pressure (P_c) if the rod-end pressure (P_r) is increased to 120 lb/in.^{2}?
19. From Figure 6.10, recalculate the extending speed of the cylinder when the rod is increased to 3 in.^{2} and the retracting speed is adjusted to 100 in./min.
20. From Figure 6.11, recalculate the pressure readings of all gauges for a new pump pressure setting of 900 lb/in.^{2}.
21. From Figure 6.15, how much flow will be bypassed to tank by the compensator spool if the throttle is set at 7 gal/min? Give your answers in (A) gal/min, (B) in.^{3}/min, and (C) cm^3/min.

7
Hydraulic Pumps

Hydraulic pumps convert mechanical power (primarily rotational power from a motor or engine) to hydraulic power. Mechanical rotational power is the product of torque and speed, whereas hydraulic power is pressure times flow. Either flow or pressure can be fixed by the design of the pump while the other parameter is allowed to swing with the load. In other words, when the flow from the pump is fixed, the pressure goes up as the load restriction is increased; and when the pump delivers fixed pressure, the flow goes down with an increase in load restriction.

7.1   PUMP CHARACTERISTICS

Some hydraulic pumps, such as the centrifugal types, produce pressure proportional to the speed of the rotor. But most pumps used in fluid power systems are of the positive displacement category. That is, they produce flow proportional to their displacement (cubic inches per revolution, in.^{3}/rev) and the speed of their rotors (revolutions per minute, rev/min).

Basic Pumping Action

Figure 1.7 of Chapter 1 is reproduced here as Figure 7.1 so we can review basic pumping action. This circuit is the same as that of Figure 1.7 except for a little more detail in the dimensions and a stop mounted on the piston rod to limit the forward stroke of the pump by two inches.

The tank is mounted 1 ft below the pump inlet. The fluid must be lifted 1 ft and pushed through the check valve V_1 and the pipes in order to get into the pump. This will require slightly more than 1.4 lb/in.^{2} (0.4 lb/in.^{2} to lift the fluid 1 ft plus 1 lb/in.^{2} to overcome the spring of V_1, plus a little more to push the oil through the pipes and cylinder port).

The tank of Figure 7.1, like the mercury reservoir of the barometer in Figure 1.8, has the atmospheric pressure of about 14.5 lb/in.^{2} (absolute) pushing down on the surface of the oil. All we have to do is create a sufficient partial vacuum in the pumping chamber

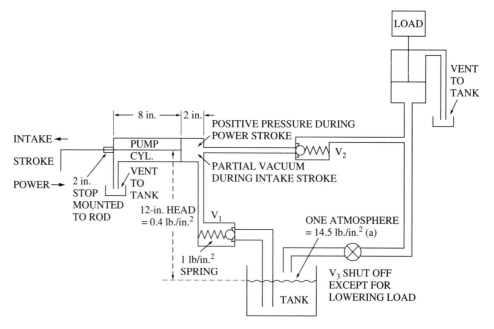

FIGURE 7.1
Basic pumping action.

and the atmospheric pressure will push the fluid up the 1-ft distance, through the check valve and the pipes, into the pump.

*Note: The atmospheric pressure at sea level with ideal weather conditions is 14.7 lb/in.²
absolute (a). However 14.5 lb/in.²(a) is assumed to be nominal and is used as a unit of
pressure in the metric system. One bar = one atmosphere = 100 kPa = 14.5 lb/in.²(a).
(Reference Chapter 2, Table 2.2)*

Any pressure generated inside the pump that is lower than atmospheric pressure (14.5 lb/in.² absolute) is in fact a partial vacuum.

We create this partial vacuum in the pump by pulling on the pump handle, expanding the volume capacity of the pumping chamber. When the pull on the handle is sufficient to lower the pressure in the pumping chamber by slightly more than 1.4 lb/in.² (or to approximately 13 lb/in.² absolute) the atmospheric pressure (14.5 lb/in.² absolute) will push the fluid through V_1, lift it 1 ft, and push it into the pump.

Note: When the volume capacity of a confined fluid is expanded, *the pressure of the fluid
will be reduced. Boyle's law states that $P_1V_1 = P_2V_2$, where P_1 is the original pressure
and V_1 the original volume, and P_2 is the reduced pressure after volume V_2 has been
increased.*

*To lower pressure below the atmospheric level is to produce a vacuum. To prevent
excessive cavitation, however, the pump inlet should be limited to a vacuum of about −5 in.
of mercury (about −2.5 lb/in.²) or an absolute pressure of not less than 12 lb/in.²(a).*

■ | **EXAMPLE 7.1: Inlet Pressure**

What will be the inlet pressure of the pump in Figure 7.1 if the pump handle is retracted 0.5 in. from the 2-in. stop prior to the start of flow? (*Hint:* Use Boyle's law.)

Solution:

$$P_1 V_1 = P_2 V_2 \text{ and } P_2 = \frac{P_1 V_1}{V_2}$$

$$= \frac{14.5 \text{ psia} \times 2}{2.5}$$

$$= \textbf{11.6 lb/in.}^2\textbf{(a)}$$ ■

Note: When solving ratios and proportions—as in Boyle's law—it is quite permissible to leave out the dimensions of the parameters. The use of dimensions could add unnecessary bulk to the solution.

Units of Displacement

Pump displacement is the volume of fluid discharged when the pump is either rotated one revolution or one radian. Units of pump displacement are:

1. in.^{3}/rev or cm^3/rev
2. in.^{3}/rad or cm^3/rad where 1 revolution = 6.28 radians and rev/(6.28 rad) = 1 and (6.28 rad)/rev = 1 (from Table 2.2).

Fixed Displacement

The displacement of some positive-type pumps is fixed, and since most electric motors and gasoline engines used to drive pumps have fixed speeds, the flow from this type of pump is also fixed at:

$$\text{Flow} = \text{Displacement } \frac{\text{in.}^3}{\text{rev}} \times \text{Speed } \frac{\text{rev}}{\text{min}} = \textbf{in.}^3\textbf{/min}$$

Variable Displacement

Some positive displacement pumps, however, are designed with a variable displacement. Some are manually adjusted but many change automatically with flow demand and are called pressure-compensated pumps. The pump pressure is fixed, but may be set by an external adjustment.

Note: Variable displacement means that the volume of fluid per revolution (in.^{3}/rev) is adjustable, either manually or automatically—as in a pressure-compensated pump. Positive displacement means that the pump rotor does not slip. Once the displacement is set, the pump will produce a given flow for that displacement and speed.

Two precautions are in order:

1. The fixed-flow pump requires a relief valve, or some other device, to handle any excess flow not needed by the load.
2. To obtain a fixed pressure from a compensated pump, the system must have enough restriction to keep the flow demand of the pump below its rated capacity.

Pump Efficiency

The general definition of % efficiency is (Output/Input) × 100. The 100 changes it to percent. Hydraulic pumps have various efficiencies:

$$\% \text{ Mechanical efficiency} = \frac{\text{Theoretical torque to drive}}{\text{Actual torque to drive}} \times 100$$

$$\% \text{ Volumetric efficiency} = \frac{\text{Actual flow rate output}}{\text{Theoretical flow rate output}} \times 100$$

$$\% \text{ Overall efficiency} = \text{Mechanical efficiency} \times \text{Volumetric efficiency} \times 100$$

$$\text{or } \% \text{ Overall efficiency} = \frac{\text{Output horsepower}}{\text{Input horsepower}} \times 100$$

7.2   TYPES OF PUMPS

Of all the types of positive-displacement pumps, the most popular are the gear, vane, and piston pumps. These three types are covered in this chapter.

Gear Pumps

Gear pumps employ various types of rotating gears meshed in housings in such a way as to pump oil or other fluids.

External Gear Pumps

The so-called "external gear" pump uses two standard spur gears as shown in Figure 7.2. They are sometimes called "external" gears to distinguish them from "internal" ring gears where the teeth are projected inward.

The two gears mesh inside a sealed housing, with one gear being driven by a shaft and the other rotated in the opposite direction by the first. As the shaft rotates, the two gears unmesh at the inlet and mesh again at the output.

The volume capacity in the inlet of the pumping chamber is expanded as the gear teeth unmesh. This produces a partial vacuum, allowing the atmospheric pressure to push fluid into the pump. As the gears rotate, fluid will be carried around the housing— in both directions—between the gear teeth. The volume capacity in the outlet of the pumping chamber is compressed as the gear teeth mesh, pushing the fluid out of the pump.

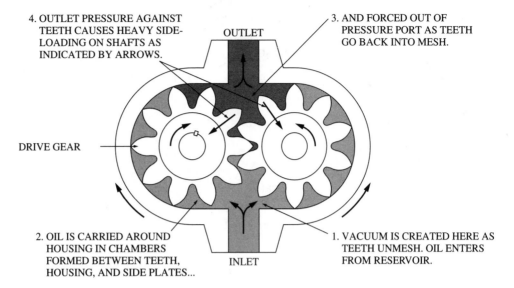

4. OUTLET PRESSURE AGAINST TEETH CAUSES HEAVY SIDE-LOADING ON SHAFTS AS INDICATED BY ARROWS.

OUTLET

3. AND FORCED OUT OF PRESSURE PORT AS TEETH GO BACK INTO MESH.

DRIVE GEAR

2. OIL IS CARRIED AROUND HOUSING IN CHAMBERS FORMED BETWEEN TEETH, HOUSING, AND SIDE PLATES...

INLET

1. VACUUM IS CREATED HERE AS TEETH UNMESH. OIL ENTERS FROM RESERVOIR.

FIGURE 7.2

External gear pump. (Courtesy of Vickers Inc.)

■ | **EXAMPLE 7.2: Hydraulic Horsepower (U.S. Units)**

What will be the hydraulic horsepower delivered by the pump of Figure 7.2 if it supplies 5 gal/min at a pressure of 1000 lb/in.^{2}?

Solution (Check formula in Table 2.6):

$$hp = \frac{Pressure \times Flow}{Constant}$$

$$= \frac{psi \times gal/min}{1714}$$

$$= \frac{1000 \times 5}{1714} = \textbf{2.92 hp}$$

■

■ | **EXAMPLE 7.3: Hydraulic Kilowatt Power (SI Units)**

What will be the hydraulic kilowatt power delivered by the pump of Figure 7.2 if it supplies 18 L/min at a pressure of 200 bars?

Solution (Check formula in Table 2.7):

$$kW = \frac{Pressure \times Flow}{Constant}$$

$$= \frac{bars \times L/min}{600}$$

$$= \frac{200 \times 18}{600}$$

$$= \textbf{6 kW}$$

■

FIGURE 7.3

Multiple external gear pump. (a)
Single pump. (b) Double pump.
(c) Standard graphic symbols
for fluid power diagrams.
(Courtesy of Vickers Inc.)

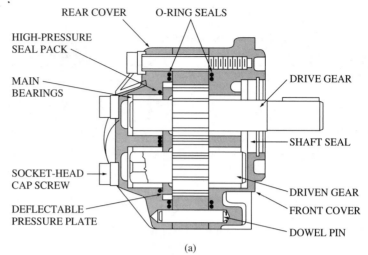

REAR COVER   O-RING SEALS

HIGH-PRESSURE
SEAL PACK

MAIN
BEARINGS

SOCKET-HEAD
CAP SCREW

DEFLECTABLE
PRESSURE PLATE

DRIVE GEAR

SHAFT SEAL

DRIVEN GEAR

FRONT COVER

DOWEL PIN

(a)

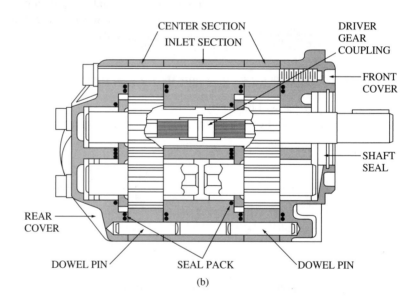

CENTER SECTION

INLET SECTION

DRIVER
GEAR
COUPLING

FRONT
COVER

SHAFT
SEAL

REAR
COVER

DOWEL PIN   SEAL PACK   DOWEL PIN

(b)

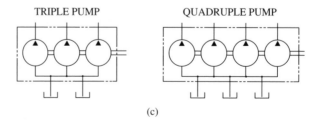

TRIPLE PUMP   QUADRUPLE PUMP

(c)

■ | **EXAMPLE 7.4: Mechanical Horsepower**

What mechanical horsepower will be required of the motor driving the pump of Example 7.2 if the overall efficiency of the pump is 80%?

Solution (Check formula of Table 2.5.): The input mechanical horsepower will be more than the hydraulic output power because the overall efficiency is only 80%. We will get a larger number if we divide by the fraction 0.80. So the answer is:

$$\text{Mechanical hp} = \frac{2.92}{0.80} = \textbf{3.65 hp}$$

■

The external pump can be designed as a single- or multiple-chamber device as shown in Figure 7.3. The multiple-chamber units can be connected in parallel to increase the pump flow capacity, or the inlets and outlets can be isolated to provide many pumps on a single shaft, some even with different fluids.

The external gear pump has an unbalanced side load on its bearings. This is caused by the high pressure at the outlet port and low pressure at the inlet both being felt by the shaft bearings. This imbalance means lower pressure ratings, slower speeds, and a lower bearing life than comparable pumps with balanced side loads.

Lobe Pump

The lobe pump (Figure 7.4) is essentially an external gear pump with three teeth per gear. It has most of the characteristics of other external gear pumps except that (1) because of its large lobes, it has more displacement per pump size and (2) since it has only three teeth per gear it has a much lower frequency of flow and pressure pulsations. It has a lower pressure rating, but is well suited for shear-sensitive fluids.

FIGURE 7.4
Lobe pump. (Courtesy of Vickers Inc.)

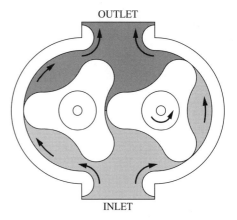

■ ### EXAMPLE 7.5: Overall Efficiency

What is the overall efficiency of the pump of Figure 7.4 if the output is 3.6 hp and the input is 4.0 hp?

Solution (Check formula in Table 2.5.):

$$\text{Overall eff.} = \frac{\text{Output hp}}{\text{Input hp}} \times 100$$

$$= \frac{3.6}{4.0} \times 100 = \mathbf{90\%}$$

■

Internal Gear Pump—Gerotor Type

One of the most common types of internal gear pumps is the gerotor shown in Figure 7.5. This pump has an external gear rotating inside an internal gear. The external gear is keyed to the shaft and has one tooth less than the internal gear.

The volume capacity of the pumping chamber is expanded as the gear teeth pass the inlet. This creates a partial vacuum at the inlet, allowing atmospheric pressure to push fluid into the pump. The fluid is carried to the pump outlet between the inner and outer gear teeth, which are in constant contact to assure an effective seal.

The gerotor pump also has a fixed displacement and an unbalanced bearing load.

■ ### EXAMPLE 7.6: Rotary Horsepower (U.S. Units)

What is the input horsepower to the pump of Figure 7.5 if the speed is 900 rev/min and the torque to the pump shaft is 100 lb-in.?

Solution (Check formula in Table 2.6.):

FIGURE 7.5
Internal gear pump—gerotor
type. (Courtesy of Vickers Inc.)

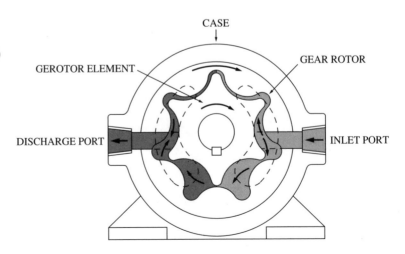

CASE

GEAR ROTOR

GEROTOR ELEMENT

DISCHARGE PORT

INLET PORT

$$\text{(Mech. rotary) input hp} = \frac{\text{Speed} \times \text{Torque}}{\text{Constant}}$$

$$= \frac{\text{rev/min} \times \text{lb–in.}}{63,025}$$

$$= \frac{900 \times 100}{63,025} = \textbf{1.428 hp}$$ ■

■ ### EXAMPLE 7.7: Rotary Kilowatt Power (SI Units)

What is the input power in kilowatts to the pump of Figure 7.5 if the speed is 900 rev/min and the torque is 40 N-m?

Solution: (Check formula in Table 2.7.):

$$\text{(Mech. rotary) input kW} = \frac{\text{Speed} \times \text{Torque}}{\text{Constant}}$$

$$\text{(Mech. rotary) kW} = \frac{\text{rev/min} \times \text{N-m}}{9550}$$

$$\text{(Mech. rotary) kW} = \frac{900 \times 40}{9550}$$

$$\text{(Mech. rotary) input kW} = \textbf{3.77 kW}$$ ■

■ ### EXAMPLE 7.8: Overall Efficiency

What is the overall efficiency of the pump in Example 7.7 if it delivers 15 L/min at a pressure of 200 bars?

Solution: (Check formulas in Table 2.7.):

$$\text{Output kW} = \frac{200 \text{ bars} \times 10 \text{ L/min}}{600}$$

$$= \textbf{3.33 kW}$$

$$\% \text{ Overall eff.} = \frac{\text{Output power}}{\text{Input power}} \times 100$$

$$= \frac{3.33}{3.77} \times 100 = \textbf{88\%}$$ ■

Internal Gear Pump—Crescent Type

The crescent-type internal gear pump shown in Figure 7.6 also has an external gear rotating inside an internal gear. The unmeshing of gears at the inlet creates a partial vacuum, allowing atmospheric pressure to push fluid into the pump. A crescent seal is machined into the housing to seal the outlet from the inlet ports. This pump also has fixed displacement, and is available in single or multiple stages.

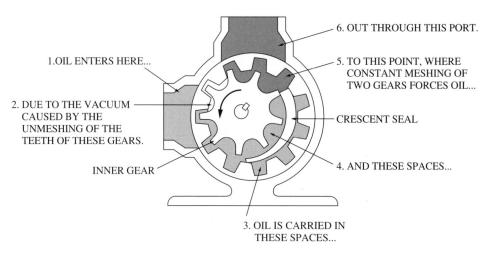

6. OUT THROUGH THIS PORT.

1. OIL ENTERS HERE...

5. TO THIS POINT, WHERE
CONSTANT MESHING OF
TWO GEARS FORCES OIL...

2. DUE TO THE VACUUM
CAUSED BY THE
UNMESHING OF THE
TEETH OF THESE GEARS.

CRESCENT SEAL

INNER GEAR

4. AND THESE SPACES...

3. OIL IS CARRIED IN
THESE SPACES...

FIGURE 7.6
Internal crescent gear pump.

EXAMPLE 7.9: System Efficiency

Calculate the following efficiencies for an assumed pumping system using the pump of Figure 7.6:

    A. Electric drive motor efficiency
    B. Hydraulic pump efficiency
    C. System overall efficiency

Use the following inputs and outputs:

1. Motor input power = 3.73 kW; motor output power = 4.5 hp
2. Pump input power = 4.5 hp; pump output flow = 7 gal/min; pump output pressure = 1000 lb/in.^{2}

Check formulas and conversions in Tables 2.1, 2.5, and 2.6.

Solution A: Electric motor eff. $= \dfrac{\text{Output hp}}{\text{Input hp}} \times 100$

$$= \frac{4.5 \text{ hp}}{3.73 \text{ kW} \times (\text{hp}/0.746 \text{ kw})} \times 100$$

$$= \frac{4.5}{5} \times 100 = \mathbf{90\%}$$

Solution B: Pump overall eff. $= \dfrac{\text{Pump output hp}}{\text{Pump input hp}} \times 100$

Pump output horsepower must first be calculated:

$$\text{Pump output hp} = \frac{7 \text{ gal/min} \times 1000 \text{ psi}}{1714} = \mathbf{4.08 \text{ hp}}$$

Then,

$$\text{Pump overall eff.} = \frac{4.08}{4.5} \times 100 = \textbf{90.6\%}$$

Solution C:

$$\text{System overall eff.} = \frac{\text{Pump output hp}}{\text{Input (electric motor) hp}} \times 100$$

$$\text{Electric mtr. input hp} = 3.73 \ \cancel{\text{kW}} \times \frac{\text{hp}}{0.746 \ \cancel{\text{kW}}} = \textbf{5 hp}$$

$$\text{System overall eff.} = \frac{4.08}{5} \times 100 = \textbf{81.6\%} \text{ or}$$

$$\text{System overall eff.} = \text{Motor overall eff.} \times \text{Pump overall eff.}$$
$$= 0.90 \times 0.906 \times 100$$
$$= \textbf{81.54\%} \text{ (close enough for most practical purposes)} \quad \blacksquare$$

Unbalanced Vane Pump

Figure 7.7 shows an unbalanced vane pump. Vanes are placed in slots milled into the rotor, which is placed off-center inside a cam ring.

When the shaft turns the rotor, the centrifugal force will push the vanes out against the cam ring, providing a hydraulic seal. The extension of the vanes as they move through the inlet indicates an expansion of volume capacity that allows the atmospheric pressure to push fluid into the pump. The vanes will carry the fluid around to the outlet, where their retraction causes the fluid to be expelled.

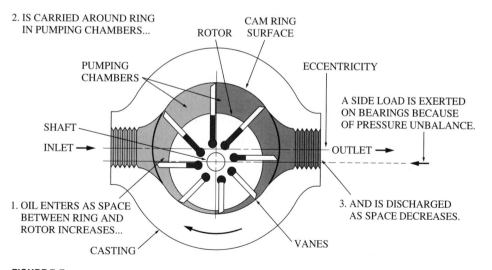

FIGURE 7.7
Unbalanced vane pump. (Courtesy of Vickers Inc.)

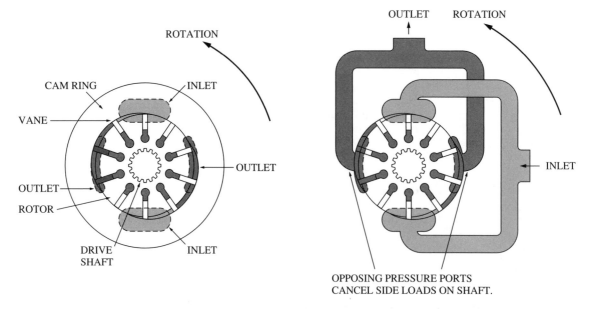

FIGURE 7.8
Balanced vane pump. (Courtesy of Vickers Inc.)

The vane pump is a good, quiet, and economical unit. The major complaint about the unbalanced pump of Figure 7.7 can be cured by the balanced design of Figure 7.8.

Balanced Vane Pump

The balanced vane pump shown in Figure 7.8 functions the same as the unbalanced design, except that the cam ring is elliptical and there are two internally connected inlet ports and two internally connected outlet ports.

Now, with this design, the two high-pressure outlet ports are located 180 degrees apart. *This port distribution produces equal and opposite side loads on the bearings that completely cancel each other.*

The balanced vane design was partly responsible for Harry Vickers' big success in the early 1930s. The bearing lifetimes of the balanced vane pumps and motors were increased 10 to 20 times over those of the unbalanced units. This also greatly increases the pressure and speed ratings of these units.

The balanced vane pumps and motors lend themselves to much versatility of design and application. There are square pumps, round pumps, reversible pumps, pumps with spring-loaded pressure plates, high-performance pumps, balanced variable vane pumps, and pumps with replaceable cartridges for easy and swift maintenance (see Figure 7.9). *If you need a pump or motor, consider a balanced vane design.*

Piston Pump

Reciprocating pistons make efficient high-pressure pumps (reference Figure 7.1). As the piston is retracted, which expands the volume capacity of the pump, a partial vacuum is

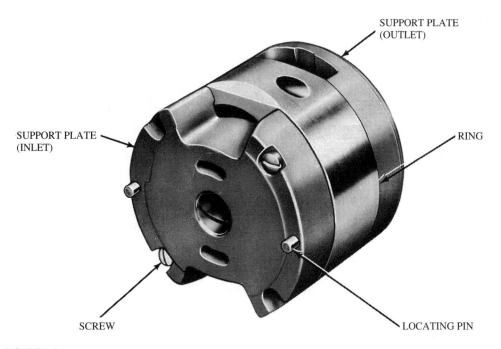

FIGURE 7.9
Preassembled cartridge. (Courtesy of Vickers Inc.)

created. This allows the atmospheric pressure to push fluid into the pump. As the piston is extended, it pushes the fluid out to the load.

The big problem is designing a high-speed reciprocating piston pump operated from a rotary motor. How do we convert rotary motion to reciprocal motion?

Radial Piston Pump

One way to convert rotary motion to reciprocal piston motion is with the radial piston pump shown in Figure 7.10. The pistons are placed in radial bores around the rotor, much

FIGURE 7.10
Radial piston pump. (Courtesy of Vickers Inc.)

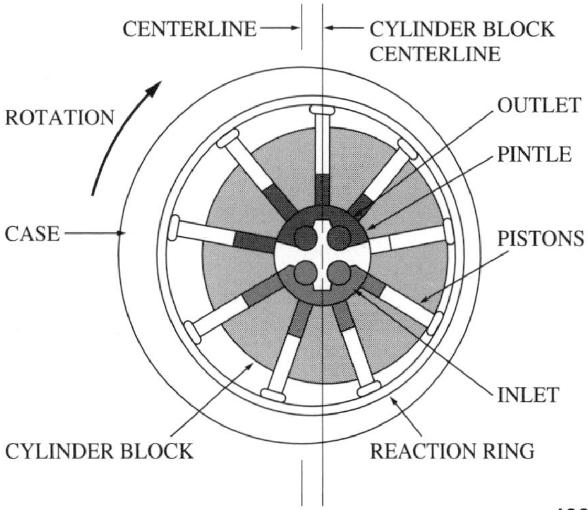

FIGURE 7.11

(a) In-line axial piston pump. (b)
Swash plate causes pistons to
reciprocate. (Courtesy of
Vickers Inc.)

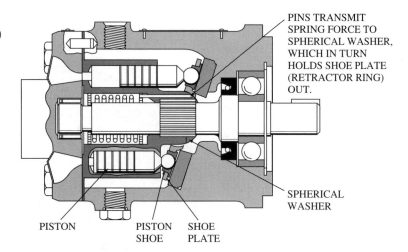

PINS TRANSMIT
SPRING FORCE TO
SPHERICAL WASHER,
WHICH IN TURN
HOLDS SHOE PLATE
(RETRACTOR RING)
OUT.

SPHERICAL
WASHER

PISTON   PISTON   SHOE
            SHOE   PLATE

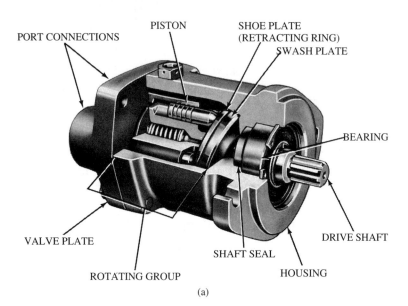

PORT CONNECTIONS

PISTON

SHOE PLATE
(RETRACTING RING)

SWASH PLATE

BEARING

VALVE PLATE

SHAFT SEAL

DRIVE SHAFT

ROTATING GROUP

HOUSING

(a)

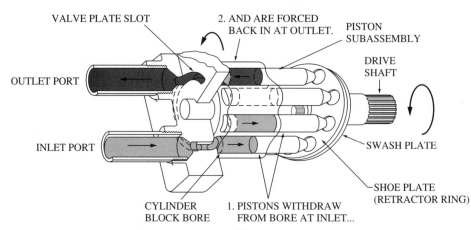

VALVE PLATE SLOT

2. AND ARE FORCED
BACK IN AT OUTLET.

PISTON
SUBASSEMBLY

DRIVE
SHAFT

OUTLET PORT

INLET PORT

SWASH PLATE

SHOE PLATE
(RETRACTOR RING)

CYLINDER
BLOCK BORE

1. PISTONS WITHDRAW
FROM BORE AT INLET...

(b)

the same as the arrangement of pistons in a radial aircraft engine. Shoes on the pistons ride on an eccentric ring, which causes the pistons to reciprocate as they rotate. A timed porting arrangement (valve plate) in the pintle connects each piston to the inlet port just as it starts to extend and to the outlet port as it starts to retract. This action is analogous to the brushes and commutator of a dc generator.

In-Line Axial Piston Pump

In the pump of Figure 7.11a, the pistons are arranged in their bores in line with the axis of the rotor. They reciprocate as they rotate because of the angled swash plate. The pistons are connected through shoes and shoe plate that bear against the swash plate. So, as the pistons follow the swash plate they reciprocate. Figure 7.11b illustrates how the swash plate causes the pistons to reciprocate.

Variable Displacement In-Line Pump

Figure 7.12a shows a variable displacement in-line pump with a pressure compensator control. The drawings of Figure 7.12b portray the variation in pump displacement as the angle of the swash plate is changed.

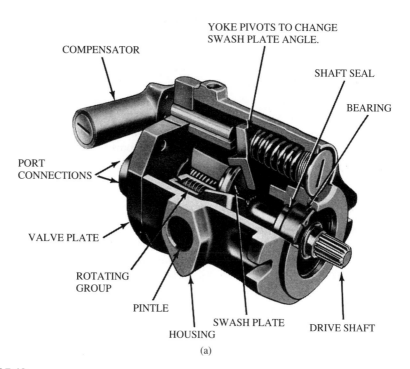

(a)

FIGURE 7.12

(a) Variable displacement of in-line piston pump. (Courtesy of Vickers Inc.)

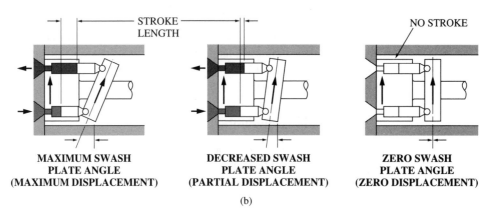

MAXIMUM SWASH
PLATE ANGLE
(MAXIMUM DISPLACEMENT)

DECREASED SWASH
PLATE ANGLE
(PARTIAL DISPLACEMENT)

ZERO SWASH
PLATE ANGLE
(ZERO DISPLACEMENT)

(b)

FIGURE 7.12 *(continued)*
(b) Variation of in-line pump displacement. (Courtesy of Vickers Inc.)

A yoke is provided to change the angle of the swash plate. The yoke has a heavy spring loading the swash plate to full-flow position. The opposite side of the yoke is mechanically connected to a cylinder, which receives fluid from the pump outlet through the compensator valve.

Figure 7.13 shows a schematic of the in-line pressure-compensated pump. Flow from the pump through the compensator valve (to tank) produces a reduced pressure (500 lb/in.^{2}) at the port connected to the cap end of the 2:1 area yoke piston. A force balance exists at the yoke where the force of the yoke spring (equal to a pressure of 500 lb/in.^{2} on the cap area of the yoke cylinder) is exactly equal to the cylinder force with 500 lb/in.^{2} supplied by the compensator valve.

When the system pressure rises to a value slightly higher than the compensator spring (1000 lb/in.^{2}), the valve will open slightly, supplying slightly more than 500 lb/in.^{2} to the yoke piston. This will slightly unbalance the yoke forces, driving the yoke down slightly toward zero, lowering the pump pressure back to 1000 lb/in.^{2}, and rebalancing the yoke forces at a pressure of 500 lb/in.^{2}.

When the system pressure is lowered, by the load, to a value slightly lower than the compensator setting of 1000 lb/in.^{2}, the valve will close slightly, supplying slightly less than 500 lb/in.^{2} to the yoke piston. This will slightly unbalance the yoke forces, allowing the yoke spring to push the yoke up to raise the pressure back to 1000 lb/in.^{2} and rebalance the yoke forces at a pressure of 500 lb/in.^{2}.

The pump outlet pressure will remain virtually constant at the compensator setting as long as the load flow is adequately restricted to maintain sufficient pressure to partially downstroke the yoke.

In other words, in order to hold a pressure, the yoke must not be bottomed out in either direction. It must remain under control, held in position by two equal forces:

FIGURE 7.13
Schematic of in-line compensated pump.

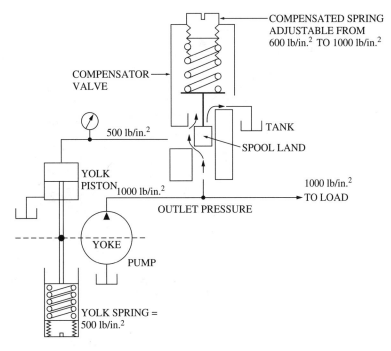

(1) the yoke spring force and (2) the force produced by the yoke piston. With these two balanced forces, any increase or decrease in pressure will automatically correct itself. If the pressure increases, the yoke piston will lower the yoke, bringing the pressure down. If the pressure goes too low, the yoke spring will raise the yoke to bring the pressure back up.

Note: Calling a valve used in this manner a compensator valve is a misnomer. It is not a compensator, it is a servo valve, and the yoke control is a pressure servo system. A servo is the only system that automatically corrects itself. The pressure compensator valve does not compensate for its mistakes; it corrects them. This point is developed more fully in Chapter 9.

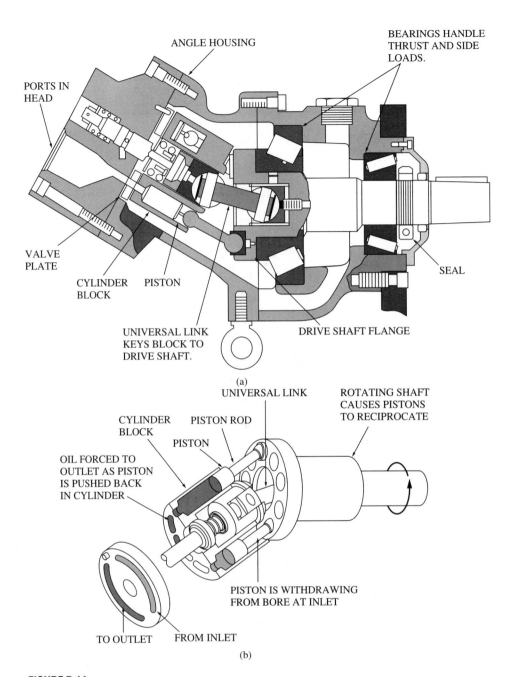

FIGURE 7.14
(a) Bent-axis piston pump. (b) Pumping action of bent-axis pump. (Courtesy of Vickers Inc.)

Bent-Axis Piston Pump

Figure 7.14a shows a bent-axis piston pump. Piston reciprocation is obtained by bending the axis of the cylinder block so that it rotates at a different angle than the drive shaft. The drive shaft turns the cylinder block through a universal link. Figure 7.14b shows the pumping action of the bent-axis pump. Figure 7.14c shows the displacement change with angle.

FIGURE 7.14 *(continued)*
(c) Displacement
changes with angle.
(Courtesy of Vickers Inc.)

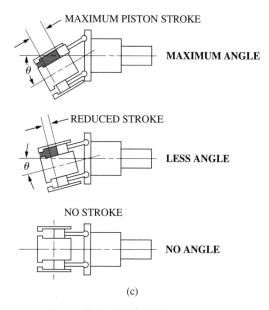

MAXIMUM PISTON STROKE

θ

MAXIMUM ANGLE

REDUCED STROKE

θ

LESS ANGLE

NO STROKE

NO ANGLE

(c)

Variable Displacement Bent-Axis Piston Pump

Figure 7.15 shows a variable displacement bent-axis piston pump. Space had to be provided inside the housing for the movement of the cylinder block assembly and yoke.

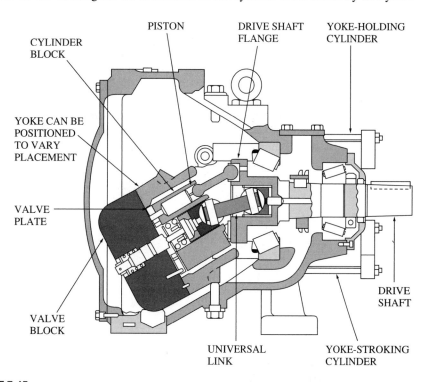

PISTON

DRIVE SHAFT
FLANGE

YOKE-HOLDING
CYLINDER

CYLINDER
BLOCK

YOKE CAN BE
POSITIONED
TO VARY
PLACEMENT

VALVE
PLATE

DRIVE
SHAFT

VALVE
BLOCK

UNIVERSAL
LINK

YOKE-STROKING
CYLINDER

FIGURE 7.15
Variable displacement bent-axis piston pump. (Courtesy of Vickers Inc.)

7.3 POWER TRANSFER

Figure 7.16 shows an electric motor driving a hydraulic pump, which powers a hydraulic motor, which moves a mechanical load. The input power to the electric motor is volts times amps divided by a constant. The input to the pump is speed times torque divided by a constant. The input to the hydraulic motor is flow times pressure divided by a constant. The input to the load is again speed times torque divided by a constant.

With 0.746 kW to the electric motor, and ignoring all losses, 1 hp is delivered to the load. But the horsepower changes form as it goes through the system. It starts out:

$$\text{EI (volts times amps)} = \cancel{746\ W} \times \frac{hp}{\cancel{746\ W}} = \textbf{1 hp}$$

into the electric motor and becomes

$$hp = \frac{(630.25\ \text{rev/min}) \times 100\ (\text{lb-in.})}{63{,}025} = \textbf{1 hp} \text{ into the hydraulic pump;}$$

$$hp = \frac{171.4\ \text{lb/in.}^2 \times 10\ \text{gal/min}}{1714} = \textbf{1 hp} \text{ into the hydraulic motor}$$

and, back again to

$$hp = \frac{(630.25\ \text{rev/min}) \times 100\ (\text{lb-in.})}{63{,}025} = \textbf{1 hp} \text{ to the load}$$

(Reference Table 2.6 of Chapter 2.) In this example both the pump and hydraulic motor are assumed to have fixed displacements. Observe that the parameters with arrows pointing to the right have their influence from left to right. That is, the voltage to the electric motor determines its speed; the speed of the pump determines its flow; the flow to the hydraulic motor determines its speed to the load.

Also observe that the parameters with arrows to the left have their influence from right to left. That is, the torque of the load determines the pressure from the pump, which determines the torque from the electric motor, which determines the current (amps) drawn from the electric power supply.

FIGURE 7.16
Power transfer.

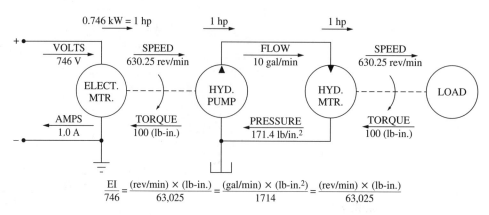

More will be said about power transfer in Chapter 9. The subject is introduced here to encourage you to "Think Systems."

7.4 PUMP APPLICATION INFORMATION IN BRIEF

1. *Gear pumps:* unbalanced, pressures to 3000 lb/in.^{2} with heavy-duty bearings, moderate leakage
2. *Vane pumps:* balanced design is the workhorse of the industry, moderate leakage
3. *Piston pumps:* balanced design, high performance with low leakage; the pistons have long lands and extensions into the cylinder bore, providing longer leakage paths and hence lower leakage.

7.5 SUMMARY

1. Hydraulic pumps convert mechanical power (mostly rotational) to hydraulic power.
2. Either flow or pressure can be fixed by the design of the pump while the other parameter is allowed to swing with the load.
3. Most pumps used in fluid power systems are of the positive displacement group.
4. Atmospheric pressure pushes the fluid into most pumps, while the pump pushes it out.
5. One bar = 14.5 lb/in.^{2} or 100 kPa is assumed to be nominal atmospheric pressure and is used as a unit of measure in the metric system.
6. When the volume capacity of a confined fluid is expanded, the pressure of the fluid will be reduced. Boyle's law states that $P_1V_1 = P_2V_2$ where P_1 is the original pressure and V_1 the original volume, and P_2 is the reduced pressure after volume V_2 has been increased.
7. When solving ratios and proportions—as in Boyle's law—it is quite permissible to leave out the dimensions of the parameters. The use of dimensions could add unnecessary bulk to the solution.
8. Pump displacement is the volume of fluid discharged when the pump is rotated one revolution or one radian.
9. Positive displacement means that once the displacement is set, the pump will produce a given flow for that displacement.
10. The fixed-flow pump requires a relief valve, or some other flow-limiting device, in order to handle the excess flow not needed by the load.
11. To obtain a fixed pressure from a compensated pump, the system must have enough restriction to keep the flow demand of the pump below its rated capacity.
12. The general definition of efficiency is

$$\frac{\text{Output}}{\text{Input}} \times 100 \text{ (to change it to percent)}$$

13. There are many types of positive-displacement pumps, the most popular being a version of a gear, vane, or piston pump. These three types are covered in this chapter.
14. Piston pumps have low leakage due to long cylinder lands, which extend deeply into the bores to provide longer leakage paths.

7.6 PROBLEMS AND QUESTIONS

1. What will be the inlet pressure of the pump in Figure 7.1 if the pump handle is retracted 0.6 in. from the 2-in. stop prior to the start of flow? (*Hint:* Use Boyle's law.)
2. What will be the hydraulic horsepower delivered by the pump of Figure 7.2 if it supplies 5 gal/min at a pressure of 1714 lb/in.^{2}?
3. What mechanical horsepower will be required of the motor, driving the pump of Problem 2, if the overall efficiency of the pump is 78%?
4. What is the overall efficiency of the pump of Figure 7.4 if the output is 4.5 hp and the input is 5.0 hp?
5. What is the input horsepower to the pump of Figure 7.5 if the speed is 1800 rev/min and the torque to the pump shaft is 250 lb-in.?
6. What is the input power in kW of the pump in Figure 7.5 if the speed is 1200 rev/min and the torque is 100 N-m?
7. What is the output power in kW of the pump of Figure 7.5 if it delivers 28 L/min at a pressure of 240 bars?
8. What is the output horsepower of the pump of Figure 7.5 if it delivers 10 gal/min at 1000 lb/in.^{2}?
9. What is the overall efficiency of the pump in Figure 7.5 if it has the input horsepower of Problem 5 and the output horsepower of Problem 8?
10. What is the overall efficiency of the pump of Figure 7.5 if it has the input power of Problem 6 and the output power of Problem 7?

Calculate the following efficiencies for an assumed pumping system using the pump of Figure 7.6:

11. Electric drive motor overall efficiency.
12. Hydraulic pump overall efficiency.
13. System overall efficiency.

Use the following inputs and outputs:

Motor input power = 1.12 kW; motor output power = 1.3 hp.

Pump input power = 1.3 hp; output flow = 1.75 gal/min; output pressure = 1000 lb/in.^{2}.

14. What will be the power delivered to the load of Figure 7.16 if the input to the electric motor is 10 hp and the electric motor, pump, and hydraulic motor each have an overall efficiency of 90%?
15. True or false?: Flow is the only parameter of a pump that can be fixed.
16. True or false?: Most pumps used in fluid power have positive displacements.
17. True or false?: Fixed-flow positive-displacement pumps have fixed flow that does not change with speed.
18. True or false?: Fluid is pushed into the pump by atmospheric pressure, not sucked in by the pump.

19. True or false?: In the SI system of standards, there are 100 bars in an atmosphere.
20. True or false?: When the volume capacity of a confined fluid is expanded, the pressure of the fluid will be increased.
21. True or false?: To prevent excessive pump cavitation, the pump inlet should be limited to an absolute pressure of not less than 12 lb/in.^{2} (a).
22. True or false?: When solving ratios and proportions—as in Boyle's law—it is quite permissible to leave out the dimensions of the parameters.
23. True or false?: Variable displacement means that the volume of fluid per revolution (in.^{3}/rev) is adjustable.
24. True or false?: The fixed-pressure pump requires a relief valve.
25. True or false?: There must be some restriction in the load in order for a pump to develop pressure.
26. True or false?: In a gear pump, the volume capacity in the inlet of the pumping chamber is expanded as the gear teeth unmesh.
27. True or false?: The external gear pump has an unbalanced side load on its bearings.
28. True or false?: The lobe pump is essentially an external gear pump with three teeth per gear.
29. True or false?: The gerotor pump has a fixed displacement and a balanced bearing load.
30. True or false?: The life of the bearings in the balanced vane pumps and motors are about the same as that of the unbalanced units.
31. True or false?: Reciprocating pistons make poor pumps.
32. True or false?: In-line piston pumps have swash plates.
33. True or false?: The pressure-compensated pump has an imbalance of forces acting on its yoke while holding a constant pressure.
34. True or false?: Increasing the load torque of Figure 7.16 will increase the amps to the electric motor.

8

Nonservo Circuits and Systems

The applications of components and principles of hydraulics covered in this text are numerous. This chapter presents a sampling of commonly used nonservo circuits. After you have learned to read and understand the function of these circuits, you should be able to read, comprehend, and draw future circuits.

8.1 POSITIONING SYSTEMS WITH VARIOUS SPEED CONTROLS

Fixed-Flow Pump and 2:1 Cylinder

Figure 8.1a shows a constant-flow pump delivering 1000 in.^{3}/min to be used for cycling the cylinder. At standby, the pump delivery is being dumped to tank through the tandem-centered three-position four-way valve (V_2). (Tandem-centered valves operate cylinders in tandem, one after the other. But because we have only one cylinder in Figure 8.1, we employ a tandem-centered valve because of its open P to T and blocked cylinder ports at center.) This dumping reduces the pressure of the pump to near zero, thus unloading the pump and electric drive motor.

The circuit drawing of Figure 8.1a, as well as the other drawings of this chapter, are arranged in a recommended pattern: (1) The sources (electric motor and pump) are on the left side and (2) the loads (cylinder or motor) on the right side of the drawing, (3) while the valves and other controls are arranged in the center from left to right—as close as possible in their order of control sequence.

Figure 8.1b shows the same circuit as Figure 8.1a except that solenoid A is now energized, shifting the upper window of the four-way valve envelope (V_2) into the circuit. This provides flow from the pump through the first flow meter, from P to A of V_2, to the cap end of the cylinder, causing the cylinder to extend. As the cylinder extends, flow from the rod end moves from B to T of V_2 through the second flow meter to tank.

Observe that the flow through the second flow meter (500 in.^{3}/min) is only half that of the first flow meter (1000 in.^{3}/min). The reason for this is that the lower flow through the second meter is from the rod end of the cylinder, which has one-half the area and hence one-half the volume per unit time (flow) of the cap end.

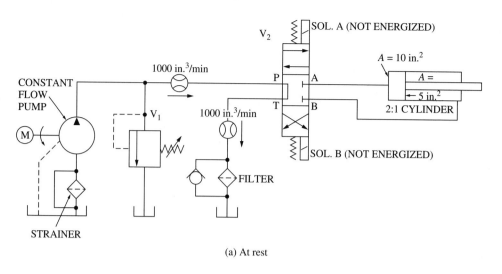

(a) At rest

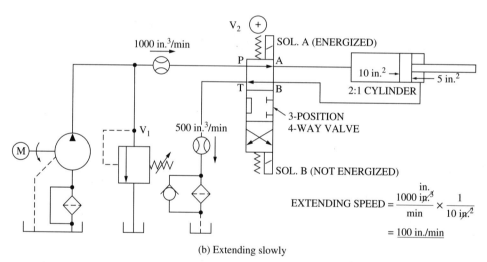

(b) Extending slowly

FIGURE 8.1
Cylinder cycling system (a) at rest, (b) extending slowly at 100 in./min, and (c) retracting faster at 200 in./min. (d) three cylinders using tandem-centered valves.

The speed of the cylinder can be calculated by using the flow of either meter:

$$\text{Speed} = \text{Flow} \times \frac{1}{\text{Area}}$$

$$\text{Extending speed} = \frac{\cancel{1000 \text{ in.}^3}\;\; ^{100 \text{ in.}}}{\text{min}} \times \frac{1}{\cancel{10} \text{ in.}^2}$$

$$= \textbf{100 in./min} \text{ or}$$

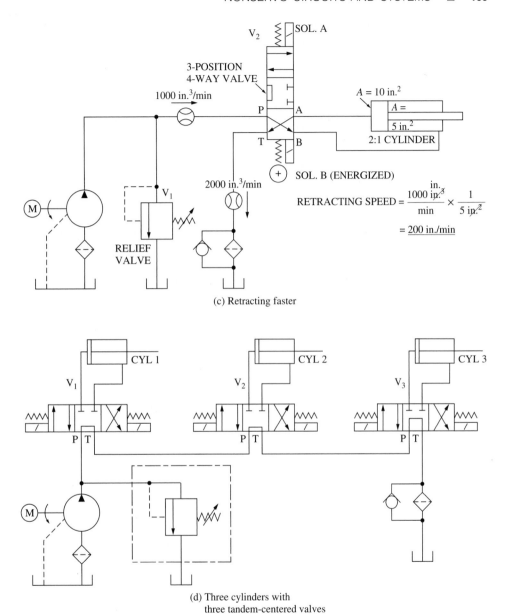

(c) Retracting faster

RETRACTING SPEED $= \dfrac{1000 \text{ in.}^3}{\text{min}} \times \dfrac{1}{5 \text{ in.}^2}$

$= \underline{200 \text{ in./min}}$

FIGURE 8.1
(continued)

(d) Three cylinders with
three tandem-centered valves

$$\text{Extending speed} = \dfrac{\overset{100 \text{ in.}}{\cancel{500 \text{ in.}^3}}}{\text{min}} \times \dfrac{1}{\cancel{5 \text{ in.}^2}}$$

$$= \textbf{100 in./min}$$

When the cylinder reaches the end of its travel, the flow of the pump will be forced to go over the relief valve (V_1) at the pressure setting of the relief valve spring. This condition will remain until solenoid A is deenergized, at which time valve V_2 will return to its tandem center, dumping the flow at near zero pressure. The circuit will now be at standby (as shown in Figure 8.1a) with very little wasted energy.

Note: Control valves with either of two center conditions may be used to dump the pump during standby:

1. Tandem center or
2. Open center.

The open-centered valve would have dumped everything to tank, including both cylinder ports. Because we want to hold the cylinder at its previous position in Figure 8.1, tandem-centered valves are used.

The circuit of Figure 8.1c is the same as that of Figure 8.1a except that (in 8.1c) solenoid B is energized instead of solenoid A shifting the lower window of the four-way valve envelope (V_2) into the circuit. This allows the pump flow of 1000 in.^{3}/min from P to B of V_2 into the rod end of the cylinder (A = 5 in.^{2}), retracting the cylinder. As the cylinder retracts, 2000 in.^{3}/min (A = 10 in.^{2}) will flow from the cap end, from A to T of V_2, and through the flow meter to tank. This will retract the cylinder at twice its extending speed as shown by the circuit of Figure 8.1b. The retracting speed can also be calculated with the use of either the rod-end flow or the cap-end flow:

$$\text{Speed} = \text{Flow} \times \frac{1}{\text{Area}}$$

$$\text{Retracting speed} = \frac{\overset{200 \text{ in.}}{\cancel{1000 \text{ in.}^3}}}{\text{min}} \times \frac{1}{\cancel{5 \text{ in.}^2}}$$

$$= \textbf{200 in./min} \text{ or}$$

$$\text{Retracting speed} = \frac{\overset{200 \text{ in.}}{\cancel{2000 \text{ in.}^3}}}{\text{min}} \times \frac{1}{\cancel{10 \text{ in.}^2}}$$

$$= \textbf{200 in./min}$$

At the end of the cycle, solenoid B should be deenergized to return V_2 to center, unloading the power from the pump and electric motor.

Cylinder Operation with Tandem-Centered Valves

Figure 8.1d shows three cylinders being operated by three identical tandem-centered valves. Observe that the pump pressure is connected to the P port of the first valve, which is internally connected to the T port at center. The T port of the first valve is externally connected to the P port of the second valve, etc., and the T port of the last valve is connected to tank. The pump flow passes through the centers of all three valves before going to tank. The pump is at standby (dumped to tank) only when all valves are at center. The A and B ports of a tandem-centered valve are blocked at center to keep the cylinder in place during standby.

The circuit of Figure 8.1d is usually used where only one cylinder is stroked at a time. For example, when cylinder 2 is stroked, pressure is applied to V_2 through the center condition of V_1 and flow to tank is through the center condition of V_3.

The advantage of this tandem circuit is that all cylinders can be at "standby" with the pump bypassed, or either cylinder can be operated without the "standby" condition of the other two valves affecting the operation. On the other hand, if the valves were connected in the conventional manner of a fixed-flow system, with all P ports connected to pressure and T ports to tank, no cylinder could be operated independently. The pump would remain bypassed until all valves were energized.

You may find it interesting to imagine what would happen if more than one valve of Figure 8.1d were operated at the same time. If all cylinders were extended in unison, the flow from the rod end of cylinder 1 would drive the cap end of 2 and the flow from the rod end of 2 would drive the cap of 3. The reverse would be true if they were all retracted at the same time.

Let's further imagine that all three cylinders are the same size and all with a 2:1 area ratio. When all cylinders are retracting, the flow out of cylinder 3 will be twice the pump flow, the flow out of cylinder 2 will be four times the pump flow, and the flow out of 1 to tank would be eight times the pump flow. *We would have a binary flow system.* When all cylinders are extending the flows would be one-half, one-fourth, and one-eighth.

Not all tandem-centered valves are used in tandem circuits. They are sometimes used when we want to dump pressure at standby but still hold the load in place. Because the tandem-centered valve does this, it is labeled as tandem-centered even though it is not used in a tandem circuit.

Storing Cylinder Rods in the Tank

When designing a hydraulic system, it is important to design the pump tank capacity large enough to store the extra fluid generated when the differential cylinders are retracted. When a differential cylinder is retracting, the volume of fluid being fed back into the tank from the cap end of the cylinder is more than the volume taken from the tank (by the pump to push the rod end). The amount of this excess volume is exactly equal to the volume of the rod of the differential cylinder.

The cylinder of Figure 8.1 has a cap-end area of 10 in.^{2} and a rod-end area of 5 in.^{2}, which means that the rod also has a cross-sectional area of 5 in.^{2}. If the cylinder has a 10-in. stroke, the volume of the rod will be:

$$\text{Volume} = \text{Area} \times \text{Length}$$
$$= 5 \text{ in.}^2 \times 10 \text{ in.} = \textbf{50 in.}^3$$

If the system of Figure 8.1 should have 10 cylinders just like the one shown, and if they were all retracted at the same time, then the tank would have to have a capacity of at least 50 in.$^3 \times 10 = \textbf{500 in.}^3$ or

$$500 \text{ in.}^3 \times \frac{\text{gal}}{231 \text{ in.}^3} = \textbf{2.165 gal}$$

Note: The tank will probably be larger than 2.165 gal, anyway, to accommodate such things as heat transfer and fluid deaeration. However, if very large cylinders, or many smaller ones, were employed, the storage of cylinder rods in the tank could be a serious design consideration.

Fixed-Flow Pump and 2:1 Regenerative Cylinder

At first glance Figure 8.2a may look like Figure 8.1a. But there is a difference. In Figure 8.2a, port A of V_2 is blocked and the rod end of the cylinder is connected back to the pump supply (through a couple of flow meters). Port B of V_2 is connected to the cap end of the cylinder. This allows the cap to be connected to pressure when solenoid B is energized and to tank when solenoid A is energized. Figure 8.2a shows the regenerative cylinder in "standby" with the pump capacity of 1000 in./min being dumped to tank through the valve's tandem center.

Note: The identification check for a regenerative cylinder circuit is that the rod end should always be connected to pressure while the cap end is alternately connected to pressure (for extension) and tank (for retraction).

Figure 8.2b shows the same system as that of Figure 8.2a except that solenoid B is energized, putting the top window of V_2 into the circuit. This connects the cap end and the rod end of the cylinder (through the flow meters) to pump pressure. The flow meters possess virtually zero restriction and have no appreciable effect on the circuit except to measure flow.

Even though the cap and rod ends of the cylinder are at the same pressure, the cylinder will extend because the cap area is larger and the same pressure will produce more extending force than retracting force. As the cylinder extends, flow from the rod end (1000 in.^{3}/min) will join the flow from the pump (also 1000 in.^{3}/min) totaling 2000 in.^{3}/min into the cap end. The extending speed of the cylinder can be calculated by using either the cap end or the rod end flow.

$$\text{Speed} = \text{Flow} \times \frac{1}{\text{Area}}$$

$$\text{Extending speed} = \frac{\overset{200 \text{ in.}}{\cancel{2000 \text{ in.}^3}}}{\text{min}} \times \frac{1}{\cancel{10 \text{ in.}^2}}$$

$$= \textbf{200 in./min} \text{ or}$$

$$\text{Extending speed} = \frac{\overset{200 \text{ in.}}{\cancel{1000 \text{ in.}^3}}}{\text{min}} \times \frac{1}{\cancel{5 \text{ in.}^2}}$$

$$= \textbf{200 in./min}$$

Figure 8.2c is the same as Figure 8.2a except that solenoid A is energized, putting the lower window of V_2 into the circuit. This connects the cap end of the cylinder to tank, which will allow the cylinder to retract in the usual fashion, with no regeneration.

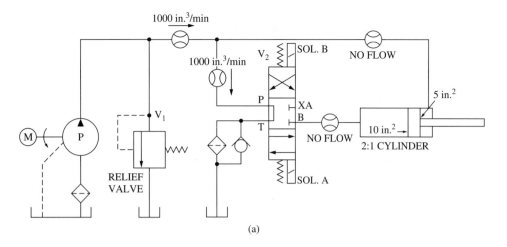

(a)

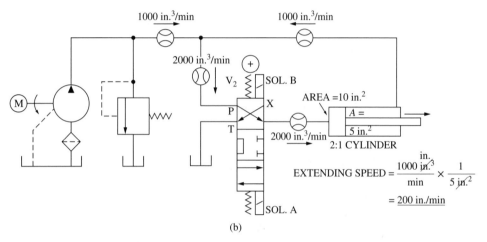

(b)

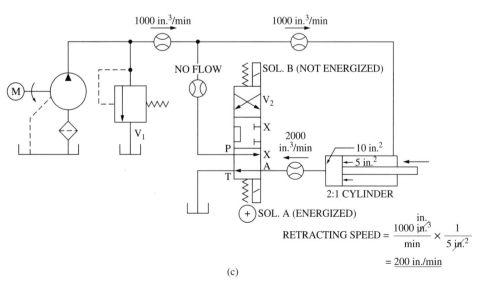

(c)

FIGURE 8.2

Regenerative cylinder (a) at rest, (b) extending at 200 in./min, and (c) retracting nonregeneratively at 200 in./min.

The retracting speed, as usual, may be calculated using either the cap-end or rod-end flow.

$$\text{Speed} = \text{Flow} \times \frac{1}{\text{Area}}$$

$$\text{Retracting speed} = \frac{200 \text{ in.}}{\frac{1000 \text{ in.}^3}{\text{min}}} \times \frac{1}{5 \text{ in.}^2}$$

$$= \textbf{200 in./min } \text{or}$$

$$\text{Retracting speed} = \frac{200 \text{ in.}}{\frac{2000 \text{ in.}^3}{\text{min}}} \times \frac{1}{10 \text{ in.}^2}$$

$$= \textbf{200 in./min } \text{(the same as extension speed)}$$

Note: The result of a 2:1 area ratio regenerative cylinder system with a fixed flow is equal extension and retraction speeds.

Visualization of Symbols

Figures 8.1a, 8.1b, 8.1c, 8.2a, 8.2b, and 8.2c were extended for instructional purposes. In the real world, there would have been two figures, 8.1 and 8.2, both shown in their center (off) position, with no separate drawings to illustrate extension and retraction flow paths.

It can become very confusing when some symbols are shifted to illustrate a portion of a cycle. For example, the symbol of a shifted sequence valve would look the same as a standard pressure-reducing valve.

For the remainder of this chapter, only one circuit will be presented for each system or subsystem illustrated. The descriptions of the circuits will be virtually the same, but you will have to visualize the valves shifting and the different flow paths caused by the shifting.

Fixed-Pressure Pump and Noncompensated Flow Valve

The circuit of Figure 8.3 provides faster extension than retraction of a cylinder. The trick is to use a constant pressure source and a noncompensated flow valve to cycle a differential cylinder. With this arrangement, the flow is limited by the restriction of the flow valve, and the differential cylinder provides pressure intensification.

A pressure-compensated pump is used to provide a fixed pressure. The three-position, four-way valve (V_2) has blocked center ports, which lead to fixed pressure at standby. While high pressure is present during standby, the flow is virtually zero. This limits the power from the pump and electric motor to near zero (another way of dumping the load).

When solenoid A is energized, the upper window of V_2 moves down to center position, allowing flow from P to A through V_2, to the cap end of the cylinder; and out the rod end through flow valve V_3, and from B to tank through V_2. The cylinder will extend at a speed set by flow valve V_3.

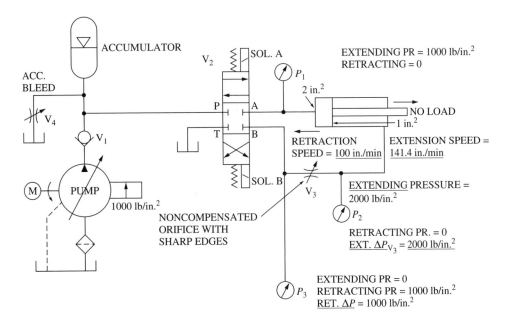

FIGURE 8.3
Fixed-pressure, noncompensated flow valve provides for faster extension.

When solenoid A is deenergized and solenoid B energized, the lower window of V_2 moves up to the center, allowing flow from P to B through V_2, through flow valve V_3, into the rod end of the cylinder, and out the cap end, through V_2 to tank. The cylinder will retract at a speed determined by the setting of flow valve V_3.

The relative extending and retracting speeds of the cylinder are determined by the square root of the relative pressure drops across the flow valve V_3. Let's assume that the flow valve is adjusted to a retracting speed of 100 in./min. Now let's look at the pressure drops across the flow valve during retraction and extension. Since the pump compensator is set at a pressure of 1000 lb/in.^{2}, the pressure drop across the flow valve is 1000 lb/in.^{2} during retraction. (No pressure exists across the cylinder during retraction because the cap end of the cylinder is connected to tank.)

During extension, the 1000 lb/in.^{2} of the pump is connected to the cap end of the cylinder. The cylinder acts as an intensifier, producing 2000 lb/in.^{2} at the rod end. This 2000 lb/in.^{2} is dropped across flow valve V_3 because the left end of V_3 is connected to tank through V_2.

$$\text{Extending speed} = 100 \text{ in./min} \times \sqrt{2000/1000}$$
$$= 100 \text{ in./min} \times \sqrt{2}$$
$$= 100 \times 1.414 \text{ in./min}$$
$$= \textbf{141.4 in./min}$$

Note: The extending speed is 1.414 times the retracting speed when a 2:1 cylinder is cycled from a fixed-pressure source and with a noncompensated bidirectional flow valve connected in the rod end of the cylinder.

The circuit of Figure 8.3 is not recommended where the cylinder must do work on the way out or back. The cylinder would slow down if it encountered a load, since the flow valve is not pressure compensated. If the cylinder must do work on the way out or back, the circuit of Figure 8.4 is recommended.

Fixed-Pressure Pump and Pressure-Compensated Flow Valves

The circuit of Figure 8.4 is similar to that of Figure 8.3 except that a meter-out, pressure-compensated flow valve is connected for each direction of movement. With this arrangement, the cylinder speed can be set for each direction, even if the cylinder encounters a load, up to the limit of the pump capacity.

When solenoid A of V_2 is energized, the upper window moves to the center, permitting flow from P to A of V_2, through the check valve of V_3 to the cap end of the 2:1 cylinder. As the cylinder extends, flow from the rod end moves through flow valve V_4, from B to T of V_2. The extending speed is determined by the adjustment of V_4 flow valve.

When solenoid A of V_2 is deenergized and solenoid B is energized, the lower window moves to the center, permitting flow from P to B of V_2, through the check valve of V_4, to the rod end of the cylinder. As the cylinder retracts, flow from the cap of the cylinder moves through flow valve V_3, and from A to T of V_2. The retracting speed is determined by the adjustment of the V_3 flow valve.

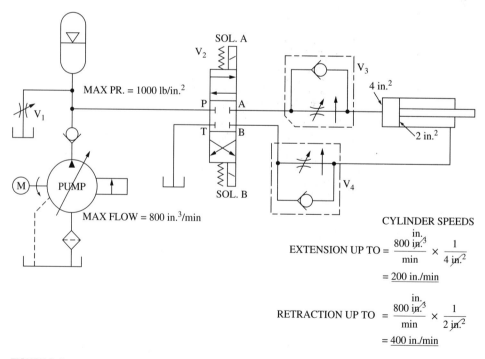

CYLINDER SPEEDS

$$\text{EXTENSION UP TO} = \frac{800 \ \text{in.}^3}{\text{min}} \times \frac{1}{4 \ \text{in.}^2}$$

$$= 200 \ \text{in./min}$$

$$\text{RETRACTION UP TO} = \frac{800 \ \text{in.}^3}{\text{min}} \times \frac{1}{2 \ \text{in.}^2}$$

$$= 400 \ \text{in./min}$$

FIGURE 8.4
Cylinder cycling with individual control of extension and retraction.

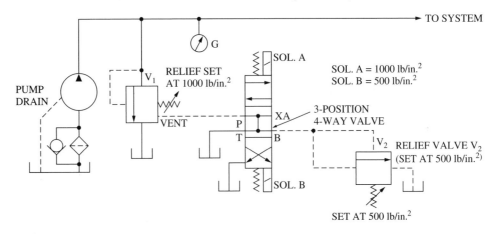

WITH ELECTRIC POWER TO SOL. A, PRESSURE = 1000 lb/in.^{2} @ G
WITH ELECTRIC POWER TO SOL. B, PRESSURE = 500 lb/in.^{2} @ G
WITH NO ELECTRIC POWER TO VALVE (AS SHOWN) = NEAR ZERO @ G

FIGURE 8.5
Two pressures plus venting.

8.2   PRESSURE MANAGEMENT AND CONTROL

Two Pressures Plus Venting

Figure 8.5 shows a fixed-flow system, which provides two maximum pressure settings plus a pump unloading mode. Two relief valves, V_1 and V_2, and one three-position open-centered four-way directional valve V_3 are used in addition to the pump.

When both solenoids of V_3 are deenergized (as shown) relief valve V_1 is vented through the open-centered V_3 valve. This reduces the system pressure to near zero.

When solenoid A is energized and solenoid B deenergized, the vent of relief valve V_1 is blocked by the upper window of V_3. This allows the system pressure to rise to a value determined by the spring setting of V_1 (1000 lb/in.^{2}) (see Table 8.1).

When solenoid B is energized and solenoid A is deenergized, the vent of V_1 is connected to the input of relief valve V_2 by the lower window of V_3. This limits the maximum system pressure to the spring setting of V_2 (500 lb/in.^{2}).

Note: Remember, to prevent coil burnout, design the electrical circuit so that both solenoids of a single valve cannot be energized at the same time (reference Chapter 4).

TABLE 8.1
Solenoid Truth Table for
Figure 8.5

Solenoid A	Solenoid B	Maximum System Pressure (lb/in.^{2})
Off	Off	0
On	Off	1000
Off	On	500
On	On	Not allowed

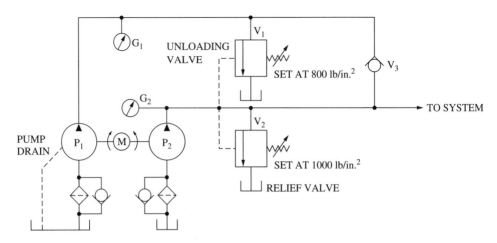

FIGURE 8.6
Two-pump system where the high pressure of one (P_2) unloads the other (P_1).

Two-Pump System

Some systems require high flow at low pressure for part of their operation and just the opposite (high pressure at low flow) for the remainder of their cycle. If just one big fixed-flow pump were used to supply both demands, much power would be wasted.

Figure 8.6 shows how lots of power can be saved by using two pumps (one high volume P_1 and one low volume P_2). Both pumps are operated from the same electric motor. Valve V_2 is a standard relief valve limiting the load pressure to 1000 lb/in.^{2}. Valve V_1 is an unloading valve with its remote sensing port also connected to the load pressure.

When the load pressure reaches 800 lb/in.^{2}, the high-flow pump (P_1) will be unloaded through V_1 allowing P_2 to supply the load, at a lower flow, up to a pressure of 1000 lb/in.^{2}. Check valve V_3 is provided to prevent the flow of P_2 from flowing through the unloading valve V_1.

A typical application of the circuit of Figure 8.6 would be to rapid advance the load, using both pumps, and hold the load under pressure using only the low-flow pump.

Accumulators and Their Application

In a pneumatic system, such as an air compressor, the fluid (air) can be stored in a tank and used at will. As more and more air is pumped into the tank, the pressure will continue to rise until the maximum safe level is reached. The air can then be released at the desired pressure level through a regulator (similar to the pressure-reducing valve shown in Chapter 5).

Since the fluid (oil) we use in most hydraulic fluid power systems is virtually incompressible, it is not so easy to store volume under pressure for future use. This problem can be solved, however, with the use of an accumulator.

The accumulator can be constructed in many forms: (1) a cylinder with a weighted piston, (2) a cylinder with a spring-loaded piston, (3) compressed gas added above the oil

FIGURE 8.7

Accumulator with an air bladder.
(Courtesy of Vickers Inc.)

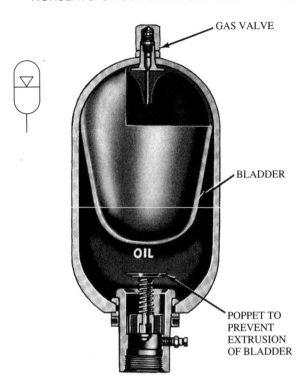

GAS VALVE

BLADDER

OIL

POPPET TO
PREVENT
EXTRUSION
OF BLADDER

in a tank, or (4) a special container with a bladder to prevent the compressed air or gas from mixing with the oil.

Figure 8.7 shows an accumulator with an air bladder. This is the most commonly used accumulator on the market. It is compact, easily used, and safe. One precaution: Avoid using combustible gases (such as oxygen) in the bladder when the hydraulic fluid is flammable (such as oil).

EXAMPLE 8.1: Accumulator Capacity

What would be the capacity of an accumulator needed to supply 5 gal of fluid to a system within the pressure range of 3000 and 2000 lb/in.^{2}, if the accumulator has a precharge of 1500 lb/in.^{2}?

Solution:

This system requires that before any hydraulic fluid is fed to the accumulator, 1500 lb/in.^{2} of gas must be stored in the bladder, inflating it to fill the entire cavity of the accumulator. Remember, Boyle's law states that as the volume of gas is reduced, the pressure goes up, or $P_1 \times V_1 = P_2 \times V_2$. As oil is pushed into the accumulator by the pump, the bladder is squeezed to a smaller size, increasing the gas pressure as the hydraulic pressure is increased.

As the pressure is increased to 3000 from 1500 lb/in.^{2}, the capacity of the bladder will be reduced by one-half. So, at 3000 lb/in.^{2} the accumulator is half filled with oil and half full of gas.

Typically, the accumulator should have the liquid capacity of twice the volume needed by the load at the highest pressure required. This means that it must store 10 gal of oil at 3000 lb/in.^{2}. Since the accumulator is only one-half full of oil at this time, it must have a total capacity (of oil and air) of 20 gal. ∎

Table 8.2 shows the conditions of the example accumulator. Observe that the bladder obeys Boyle's gas law such that the product of gas pressure and volume is equal for each volume ($P_1V_1 = P_2V_2 = P_3V_3$). The oil stored in the accumulator takes up whatever space is left over after the bladder is satisfied.

TABLE 8.2
Conditions of Example Accumulator

	V_1	V_2	V_3
Volume of gas	20 gal	15 gal	10 gal
Volume of oil	0 gal	5 gal	10 gal
Pressure	1500 lb/in.^{2}	2000 lb/in.^{2}	3000 lb/in.^{2}
Gas pressure × V	1500 × 20 = 30,000	2000 × 15 = 30,000	3000 × 10 = 30,000
Boyle's gas law	P_1V_1 =	P_2V_2 =	P_3V_3

One negative for the accumulator application is that the pressure is reduced while the fluid is being drawn from the system. Table 8.2 shows that the system pressure was reduced from 3000 to 2000 lb/in.^{2} as 5 gal of oil was taken from the accumulator. An analog of this could be that a blood donor would experience a very small reduction in blood pressure while a pint of blood is drawn from his vein.

The electrical analogy of the accumulator is the capacitor. Like the capacitor in a dc system, the accumulator smooths out ripples in pressure, pressure pulses, pressure spikes, and provides for much smoother system operation—especially where pressure-compensated pumps are used.

Electrically Unloading the Pump When Accumulator Is Charged

Figure 8.8 shows an accumulator being charged through a check valve V_1 by a standard fixed-flow pump. The relief valve V_2 is normally vented through the spring offset position of V_3.

When the pump motor is started, solenoid A is energized through the "motor start" switch and the normally closed (NC) contacts of the pressure switch PS_1. Solenoid A will shift V_3, deventing V_2 and allowing the pump to start charging the accumulator.

When the accumulator has reached the desired pressure as determined by the spring setting of PS_1, the contacts of the pressure switch will open, deenergizing solenoid A.

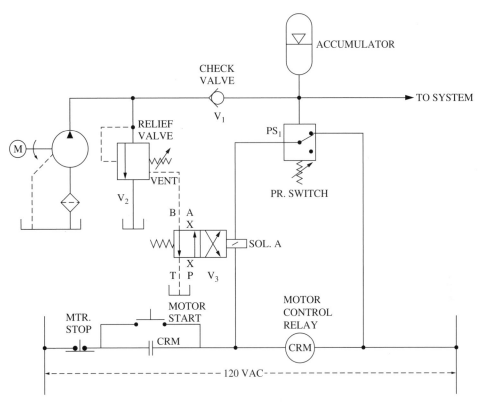

FIGURE 8.8
Pump is electrically unloaded when accumulator is charged.

When solenoid A is deenergized, V_3 will return to spring-offset, venting the relief valve and unloading the pump. Check valve V_1 is provided to prevent the fluid in the accumulator from backing up through the vented relief valve. The contacts of PS_1 will close again, allowing the accumulator to recharge, when the pressure is lowered to a preset level.

Charging Accumulator with a Pressure-Compensated Pump

If you want it simple, all you need to charge an accumulator is a pressure-compensated pump and a check valve (Figure 8.9).

When the accumulator charge reaches the compensator setting of the pump, the yoke of the pump strokes back to near zero, producing nearly zero flow. The check valve is needed to keep the fluid from the accumulator from trying to back up through the pump. Thus, the pressure-compensated pump is unloaded by reducing its flow to nearly zero instead of the pressure to near zero as in Figure 8.8.

FIGURE 8.9
Charging accumulator with a
pressure-compensated pump.

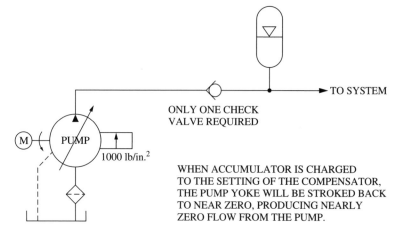

ONLY ONE CHECK
VALVE REQUIRED

WHEN ACCUMULATOR IS CHARGED
TO THE SETTING OF THE COMPENSATOR,
THE PUMP YOKE WILL BE STROKED BACK
TO NEAR ZERO, PRODUCING NEARLY
ZERO FLOW FROM THE PUMP.

Automatic Bleed-off of Accumulator

Figure 8.10 shows a circuit for automatically bleeding the stored fluid from an accumulator to tank. When the system is turned off, deenergizing solenoid A, the accumulator will discharge through the spring-offset position of V_2. A restriction connected in series with the valve will control the bleed-off rate of flow.

FIGURE 8.10
Accumulator automatic bleed-off
circuit.

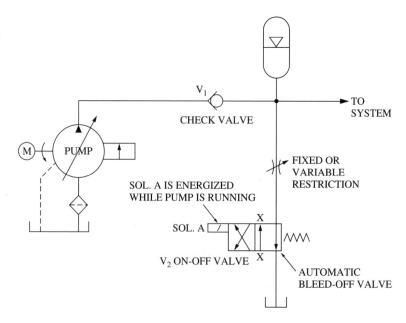

8.3 **COUNTERBALANCE, BRAKE, AND SPEED OVERRUN SYSTEMS**

For this section, reference Figures 5.6 and 5.7 of Chapter 5.

Cylinder Lift with Counterbalance

Figure 8.11 shows a circuit where a load weight can be lifted by energizing solenoid B and lowered by energizing solenoid A. However, as stated before, *never energize both solenoids at the same time.*

Without counterbalance valve V_3, the load would free fall when V_2 is shifted to center position. Even though there is some restriction of flow from tank, through the open center of V_2 to the cap end of the cylinder at this time, the load would not wait for the flow. The heavy weight would create a vacuum at the top of the cylinder allowing the load to free fall.

With the counterbalance valve installed as shown, the pressure at the rod end of the cylinder would have to build up to 800 lb/in.^{2} (spring setting of V_3) in order for the load to move down. The rod-end pressure caused by the load weight would only be:

$$\text{Pressure} = \text{Weight} \times \frac{1}{\text{Area}}$$

$$= 2000 \text{ lb} \times \frac{1}{3 \text{ in.}^2} = \textbf{667 lb/in.}^2$$

So, to move the load down, an additional pressure is needed from the pump to overcome the counterbalance valve.

When solenoid B is deenergized and solenoid A is energized, the pump will supply the additional pressure to lower the load. A pressure of 800 lb/in.^{2} − 667 lb/in.^{2} = 133 lb/in.^{2} is needed at the rod end of the cylinder. Since the cylinder has a 2:1 area ratio, the pump needs to supply only half this amount or 67 lb/in.^{2} to the cap end of the cylinder.

FIGURE 8.11
Cylinder lift circuit with counterbalance valve.

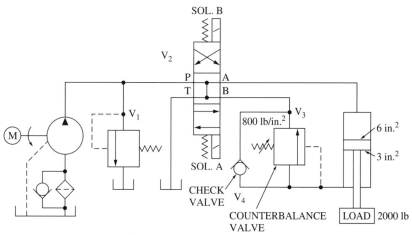

Check valve V_4 provides free flow from V_2 to the rod end of the cylinder, bypassing the counterbalance valve during the "load lift" cycle.

Motor System with Brake and Overrun Protection

Figure 8.12 shows a motor system designed to drive a flywheel or other high-inertia load. The brake and counterbalance valve (BCV), connected in the exhaust circuit of the motor, opens wide for rapid acceleration but partly closes to add restriction during overrun and braking.

The inlet pressure, which is high during acceleration, is sensed by the large control piston of the brake valve. This high inlet pressure keeps the valve wide open during acceleration.

When the load tends to overrun due to inertia, the motor inlet pressure drops and the spring of the brake valve adds some restriction to the motor exhaust. To prevent the BCV from overcorrecting the speed overrun, the exhaust port of the motor is connected to the small piston of the BCV. The small piston takes over and regulates the motor speed during deceleration.

When the manual valve (V_2) is returned to center, the load flywheel will continue to rotate, driving the motor as a pump, until the energy stored by the load inertia is dissipated. The motor will function as a pump, with its inlet port connected to tank and its outlet flow through BCV back to tank.

The small piston senses the pressure drop across the BCV valve and opens in proportion to the pressure, thus regulating the braking velocity.

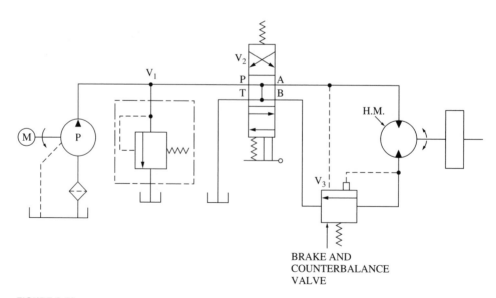

BRAKE AND
COUNTERBALANCE
VALVE

FIGURE 8.12
Motor system with brake and overrun protection.

8.4   POSITION-SENSING SYSTEMS

The systems discussed in this section sense cylinder position(s) to program another operation. The second operation may be simply to vent itself or it may cause a second cylinder to move.

One-Cycle System with Automatic Venting at End

Figure 8.13 shows a system that extends with the push of a switch, retracts automatically, and vents a relief valve at the end of its cycle.

A brief electrical circuit of the one-cycle system is shown at the bottom of Figure 8.13. When the normally open (NO) "Cycle start" push button is pushed, the coil of the control relay, 1CR (line 1), will be energized, closing all normally open contacts (lines 2 and 3) of the control relay. One set of 1CR contacts (line 3) will connect the 120 VAC to solenoid B, while the other set of contacts (line 2) provides a holding circuit through the NC contacts of the limit switch 1LS. This holding circuit keeps the coil of relay 1CR energized after the "Cycle start" push button is released.

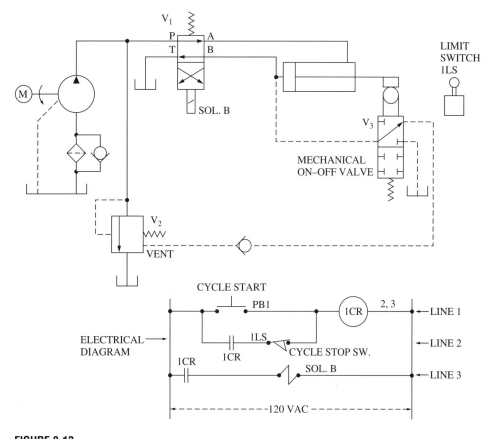

FIGURE 8.13
One-cycle system with automatic venting at end.

During standby, and until after solenoid B is energized, relief valve V_2 is vented through the upper window of V_3 and the upper window of V_1. When solenoid B is energized, the lower window of V_1 is moved up to provide flow from P to B and from A to T. This porting of V_1 will cause the cylinder to advance until limit switch 1LS is actuated, opening the holding circuit of the relay 1CR.

When the relay drops out, solenoid B is deenergized, and the valve spring returns the valve spool to its original position, porting flow from P to A and from B to T. This valve porting retracts the cylinder until the two-way mechanical on–off valve, V_3, is tripped, venting the pump relief valve V_2 and putting the system back into standby.

The check valve is added to the relief vent circuit to prevent the pump flow from backing up into the relief valve while the cylinder is extending sufficiently to release the mechanical plunger of V_3.

Regenerative Advance with Pressure-Activated Changeover

Remember that (1) a regenerative 2:1 cylinder will extend and retract at the same speed; (2) characteristically the rod end of the cylinder is connected to pressure for both directions of travel, while the cap end is connected to pressure for extension and to tank for retraction; and (3) less extending force is exerted by the regenerative cylinder than by the straight cylinder (reference Figure 8.2).

Figure 8.14 shows the circuit of a mold press using a 2:1 area cylinder connected as a regenerative circuit for advancement to the point of load contact. When the load pressure builds up to the spring setting of sequence valve V_3, the cylinder continues to advance nonregeneratively until the press bottoms out in the mold.

The three-position, four-way valve, V_2, has two stages. A 50 lb/in.^{2} check valve is connected in series with the pump, and the pilot pressure for V_2 is taken from the pump side of this valve. This circuit ensures that pilot pressure is always available, even when V_2 is in center position.

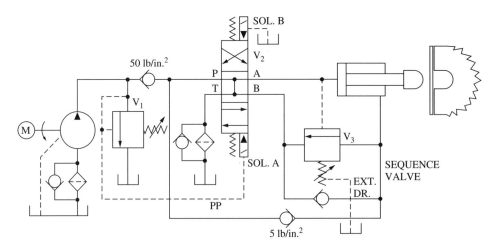

FIGURE 8.14
Regenerative advance with pressure-activated changeover.

When V_2 is centered (both solenoids deenergized), the cylinder is at rest with both ends at or near tank pressure.

When solenoid A is energized, the bottom window of V_2 moves up to the center, connecting pressure to A and B to tank. Flow from the rod end of the cylinder will pass through the 5 lb/in.^{2} check valve and join the flow from the pump, providing twice the pump flow into the cap end. Thus the cylinder "regenerates" (doubles) the pump flow into the cap end of the cylinder, providing a regenerative speed twice that of a conventional cylinder where the rod end is returned to tank.

When the load pressure builds up to the spring setting of sequence valve V_3, the rod end of the cylinder is shifted directly to tank through V_3 and V_2. The cylinder continues to advance nonregeneratively (at one-half speed, but at twice the force) until the press bottoms out in the mold. Connecting the cylinder rod end directly to tank doubles the extending force of the cylinder, providing for a better press.

Deenergizing solenoid A and energizing B will move the upper window of V_2 down to the center, connecting pressure to B and A to tank. Fluid will move through the V_3 bypass check valve into the rod end of the cylinder. The cylinder will retract normally, pushing flow out of its cap end through V_2 (from A to tank), and the cycle is complete.

Two-cylinder Sequence

When both solenoids of V_5 are deenergized (as shown in Figure 8.15), the valve is at its open center condition with both cylinders at rest. Pilot pressure is assured for two-stage valve V_5 by check valve V_6, which is placed in the tank line.

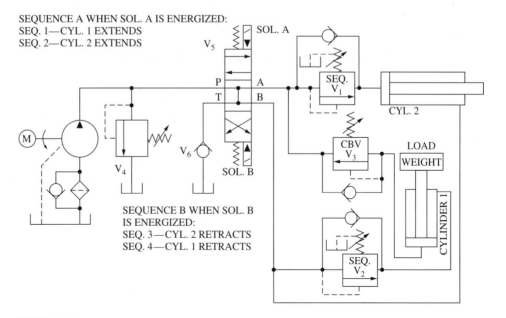

FIGURE 8.15

Two-cylinder sequence with two sequence valves and one counterbalance valve (CBV).

When solenoid A is energized, cylinder 1 extends fully (sequence 1). Then cylinder 2 extends fully (sequence 2).

When solenoid A is deenergized and solenoid B energized, cylinder 2 retracts fully (sequence 3). Then cylinder 1 retracts fully (sequence 4).

Note: Remember that when there is a choice of cylinder paths, flow will first move the cylinder that requires the least pressure (reference Chapter 6, Figure 6.1c).

When solenoid A is first energized, cylinder 1 is the first to move (extend) because it has the flow path with the lowest pressure requirement.

There are two possible flow paths:

1. Over sequence V_1 spring setting, through cylinder 2 to tank.
2. Through CBV V_3 bypass valve, cylinder, and sequence V_2 bypass valve to tank.

The system chooses path 2 because the two check valves and the load weight require less pressure than the spring setting of V_1.

After cylinder 1 fully extends, the pump pressure will build up to the spring setting of sequence valve V_1. Valve V_1 will pass flow into the cap end of cylinder 2, causing it to move to the right, where it will either fully extend or be stopped by the shifting of valve V_5.

When solenoid A is deenergized and solenoid B is energized the same type of action takes place as when solenoid A was energized. Flow again takes the path with the lowest pressure requirement—into the rod end of cylinder 2, through the cylinder, and out through the V_1 bypass check. This path requires less pressure than that necessary to overcome the spring settings of sequence V_2 and CBV V_3. So, cylinder 2 retracts first.

After cylinder 2 retracts all the way, the system pressure will build to a level that will compress the springs of V_2 and V_3. This will retract cylinder 1 and lower the load weight.

Rapid Advance to Feed Using a Deceleration Valve

Figure 8.16 shows an automatic "feed" system in which a deceleration valve is cammed to a fully open position, which is desirable when rapid advance is needed. When a slower "feed rate" is required, the deceleration valve is cammed to its closed position, forcing the cylinder exhaust flow through the adjustable flow-control valve V_3. A reverse-flow check valve, V_4, bypasses both the flow valve and the deceleration valve for a rapid return.

An electric circuit and a sequence diagram are provided in Figure 8.16. If electric circuits are new to you, call on a friend who is electrically inclined and/or check out a good electrical circuits book. If neither of the above is available to you, please be patient while I walk you through the system; we will work back and forth among the hydraulics, sequence, and electrical diagrams of Figure 8.16.

Sequence Diagram

First observe the action lines of the sequence diagram of Figure 8.16 (those with horizontal arrows: "Rapid advance," "Feed," and "Rapid return"). The action lines are interrupted by vertical lines that show the altered action of the cylinder. The cycle is started

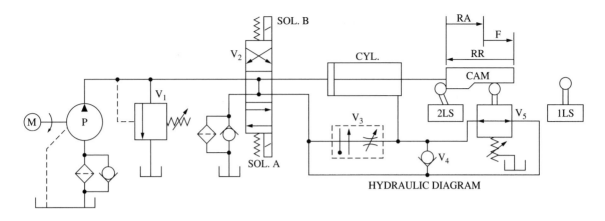

HYDRAULIC DIAGRAM

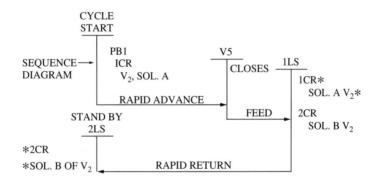

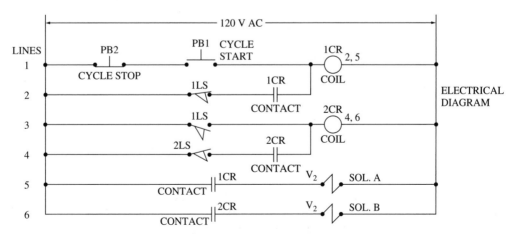

FIGURE 8.16
Rapid advance to feed using a deceleration valve.

by PB1 (push button number one), which energizes 1CR (control relay number one), which energizes solenoid A of V_2. Valve V_5 shifts the cylinder to "Feed"; limit switch 1LS stops the "Feed" and starts "Rapid return"; and 2LS stops the cylinder and returns V_2 to center, dumping the pump pressure at the end of the cycle.

Observe the asterisks (*) after 1CR and solenoid A at the end of "Feed" and in front of 2CR and solenoid B of V_2 after the "Rapid return." These asterisks tell us that these items have been turned off by the limit switches, 1LS and 2LS.

Electrical Diagram

The small circles to the right of the drawing are the relay coils. The vertical parallel lines represent the contacts of the relays. The horizontal lines of the drawing are numbered and the line number for each set of contacts is shown to the right of the relay coils. The sequence of events is as follows: Valve solenoids are turned on by relay contacts, which are turned on by relay coils, which are turned on by limit switches or push buttons.

When the "Cycle start" push button PB1 is pushed, the coil of 1CR (number one control relay) is connected across the 120 VAC (line 1). The NO relay contacts on line 2 will close, latching the 1CR relay through the NC contacts of 1LS (also on line 2). The NO 1CR contacts on line 5 will close, energizing solenoid A of V_2 and causing the cylinder to extend.

Note: This latching means that PB1 may be released and the coil of the 1CR relay will remain actuated through its own contacts (1CR) and the NC contacts of 1LS on line 2.

The cylinder will extend until V_5 is turned off by the in-line cam attached to the cylinder. This forces the cylinder discharge flow to pass through flow control valve V_3, which sets the feed rate of the cylinder.

When 1LS is tripped (at the end of the cylinder travel), 1CR is turned off, stopping the advance of the cylinder, and 2CR is turned on, providing rapid return. During rapid return, relay 2CR is latched in through 2LS (which is NC at the time). At the end of the cycle, 2LS is tripped, dropping out 2CR, turning off solenoid B, placing the system on "standby" operation.

Note: The official designation for 2LS is NCHO (normally closed, held open as shown).

Rapid Advance to Feed Using a Solenoid-Operated On–Off Valve

Figure 8.17 shows a "Rapid advance to feed" system that uses a limit switch actuated, solenoid-operated, on–off valve for switching to "feed speed." The system is the same as Figure 8.16 except that an on–off solenoid valve (V_5) is used instead of the deceleration valve. The sequence and electrical circuits are the same as in Figure 8.16 except for the addition of solenoid A of V_5, which is actuated by 3LS (line 7).

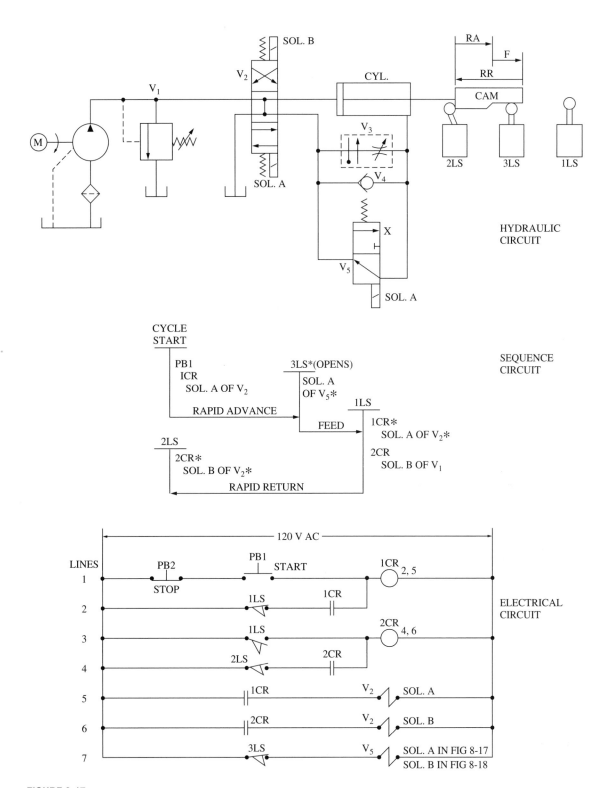

FIGURE 8.17
Rapid advance to feed using a solenoid-operated on–off valve.

Limit switch 3LS is normally closed (NC) and is cammed open when we want feed speed. Valve V_5 is shown in the "on" condition in the hydraulic diagram as well as in the electrical diagram. It is important that both diagrams agree with the physical drawing of the limit switches.

Valve V_5 is relatively fail-safe as used in this system. If the solenoid of V_5 should fail, the spring will shift the valve spool to feed speed mode even if the control logic is calling for rapid advance. Because the feed speed is slower, it should also be safer.

Rapid Advance to Feed Using a PO Check Valve

Figure 8.18 shows another system for obtaining "Rapid advance to feed" and "Rapid return." This circuit provides the same outcome as the systems of Figures 8.16 and 8.17. The only difference is that a pressure-operated (PO) check valve, V_4, is used, instead of the deceleration and on–off valves of the previous circuits, to bypass the flow valve. The PO check is turned on by on–off valve V_5.

Once again, because the control system (cam and limit switches) shows the system to be in "Rapid" mode, valve V_5 is shown in the "on" position to provide rapid speed in either direction. This means that solenoid B of V_5 is energized, providing pump pressure to the pilot port of the PO check V_4.

The sequence and electrical drawings of Figure 8.17 apply to the system of Figure 8.18 except that solenoid B of V_5 is used in Figure 8.18 instead of solenoid A.

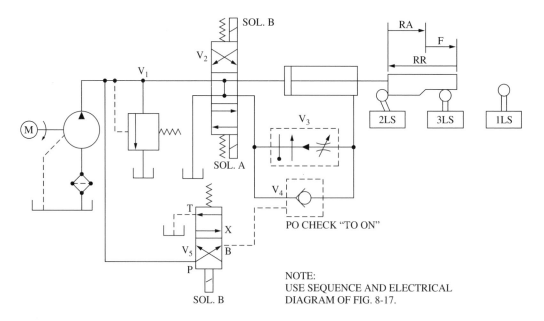

FIGURE 8.18
Rapid advance to feed using a PO check valve.

Clamp and Work System with Controlled Clamping Pressure

Figure 8.19 shows another clamp and work system where limit switches, relays, and solenoid-operated valves are used along with the hydraulic circuit. This circuit is a little more complex than the systems of Figures 8.16, 8.17, or 8.18 because it requires a bit more automation. As before, if the three previous figures represent your only exposure to electric circuits, this may be the time to call on a friend who is electrically inclined, and/or check out a good electrical circuits book. If these are not available to you, then, as before, I will walk you through the hydraulics, sequence, and electrical diagrams of Figure 8.19. There is some repetition here of the explanation of Figure 8.16. But bear with me on this, because repetition may be just the ticket in this instance.

First observe the action lines of the sequence diagram of Figure 8.19b (those with horizontal arrows). When the "Cycle start" button is pushed, the clamp cylinder extends, clamping the load. When the clamp is in, the work cylinder extends automatically. After the work is completed, the work cylinder retracts automatically. After the work cylinder retracts, the clamp cylinder retracts, automatically unclamping the load, and the cycle is complete.

Observe the finer points of the sequence diagram. Notice that the items listed in the diagram are indented to show the order of operation. PB1 activated 1CR (number one control relay), which activated solenoid A of V_4. At the same time, PB1 activated 2CR (number two control relay), which activated solenoid A of V_5. Also, observe the items hanging on dashed lines below the action lines in Figure 8.19b, for example, 2LS just after the start of the clamp extension. This means that 2LS dropped out of its held-open position just after the clamp started to extend. 3LS hangs below the "Work cylinder extend" action line. This means that 3LS dropped out of its held-closed position just after the work cylinder started to extend.

Now let's walk through the entire cycle of Figure 8.19. "Cycle start" button PB1 (lines 2 and 3) is pushed, energizing the coils of relays 1CR and 2CR.

Note: Observe that the horizontal lines of the electrical drawing are numbered on the left side of the drawing. The numbers to the right of the relay coils identify the lines on the drawing where the contacts of that particular relay may be found. Relay contacts shown with an underline (after the 4CR coil on line 8 and NC contacts of 4CR on line 6) are normally closed. Components in the sequence diagram marked with an asterisk (*) were disengaged at that point instead of being turned on (for example, solenoid V5 A * means that solenoid A of V_5 was turned off (16) after 4CR was energized (15).

The 1CR contacts on line 1 are a set of "holding contacts," which allows the relay coil of 1CR to remain energized after PB1 is released. Since this set of 1CR contacts is in series with 1LS (NC) (number one limit switch normally closed), 1CR is "latched" to 1LS, and will unlatch (fall out) when 1LS is tripped at the end of the work cycle.

But, back to the main cycle; another set of 1CR contacts is located on line 9 of the diagram. This set turns on solenoid A of V_4, but nothing yet happens in this branch of the hydraulic circuit because sequence valve V_2 has not yet shifted. The second set of PB1

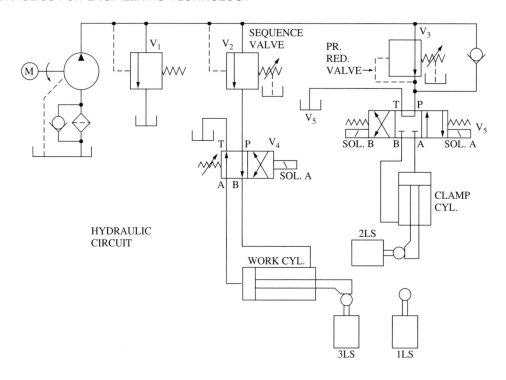

HYDRAULIC
CIRCUIT

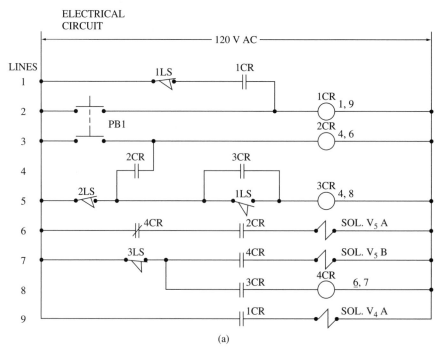

FIGURE 8.19
Clamp and work system with controlled clamping pressure.

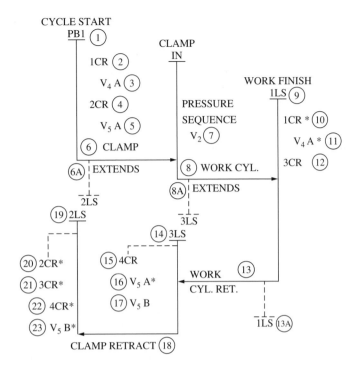

(b)

SEQUENCE OF EVENTS

① PB1 ENERGIZES

② 1CR WHICH ENERGIZES

③ V_4 A WHICH READIES WORK CYL.

① PB1 ALSO ENERGIZES

④ 2CR WHICH ENERGIZES

⑤ V_5 A WHICH

⑥ EXTENDS CLAMPS AND JUST AFTER CLAMP CYL. MOVES

⑥A 2LS CLOSES

⑦ PRESSURE SEQ. V_2 OPENS, ALLOWING

⑧ THE WORK CYL. TO EXTEND AND JUST AFTER,

⑧A 3LS OPENS

⑨ 1LS DROPS

⑩ 1CR WHICH DROPS

⑪ V_4 A CAUSING THE

⑬ WORK CYL. TO RET.

* ⑨⑫ 1LS ALSO ENERGIZES 3CR

⑬A 1LS DROPS TO NORM

⑭ 3LS IS TRIPPED, WHICH

⑮ ENERGIZES 4CR WHICH

⑯ DROPS SOL. V_5 A AND

⑰ ENERGIZES V_5 B

⑱ WHICH RETRACTS CLAMP

⑲ OPENING 2LS, WHICH

⑳ DROPPED 2CR AND

㉑ 3CR WHICH DROPPED

㉒ 4CR WHICH DROPPED

㉓ V_5 B, PUTTING THE SYSTEM IN STANDBY

FIGURE 8.19

(continued)

contacts located on line 3 turns on the coil of the 2CR relay. The push button must be held in for a fraction of a second to ensure that limit switch 2LS (number two limit switch with normally closed, held-open contacts, NCHO) (line 5) is released back to its NC position. The contacts of 2CR (line 6), in series with the NC contacts of 4CR, turn on solenoid A of V_5. This shifts the spool of V_5 to the left, porting fluid to the cap end of the clamp cylinder, extending the cylinder. The relay coil of 2CR is now latched through the 2LS contacts on line 5 and its own contacts on line 4.

When the clamp cylinder bottoms out, a low adjustable pressure is provided to the clamp by pressure-reducing valve V_3. The pump pressure will continue to build until it reaches the spring setting of sequence valve V_2, extending the work cylinder. When the work cylinder moves a short distance, the NOHC contacts of 3LS (line 7) will be released back to their NO position.

When the work cylinder reaches the end of its travel, 1LS will be tripped. The NC contacts on line 1 will open, dropping 1CR and solenoid A of V_4. This will retract the work cylinder.

Another action will take place at the end of the work cylinder extension. A pair of NO contacts on 1LS (line 5) will close momentarily, energizing relay 3CR. The coil of relay 3CR will be latched by its own contacts located on line 4 through 2LS. Relay 3CR is known as a *sequence relay*. It is used here to allow the "Work cylinder" to initiate the retraction of the clamp cylinder after the work cylinder has moved forward and tripped 1LS.

When the work cylinder is fully retracted, limit switch number three (3LS) on line 7 will be actuated and will perform the following functions: Relay 4CR will be actuated (line 8) through the previously mentioned relay contacts 3CR. This will deenergize solenoid A of V_5 (line 6) and energize solenoid B (line 7). These two actions will reverse the clamp cylinder, and open the clamp.

When the clamp is fully open, limit switch 2LS will return to its held-open position, stopping everything. Relays 2CR, 3CR, and 4CR along with solenoid B of V_5 will all be deenergized, and the open center of V_5 will reduce the pump pressure to a minimum for standby.

8.5   MOTOR CONTROL CIRCUITS

Open-Circuit Motor Drive

Figure 8.20 shows an open-circuit, manually operated motor drive. The system is considered to have an open circuit because the flow from the motor is returned to tank and not to the pump inlet as in a closed-circuit system.

The system functions as a fluid drive shaft. The speed of the motor is proportional to the displacement of the pump and is inversely proportional to the displacement of the motor. This type of circuit could incorporate a flow valve as a speed control or a variable displacement motor could be used to vary the speed.

The brake and overrun valve of Figure 8.12 could be added to the circuit of Figure 8.20 if a flywheel or other load inertia is present.

FIGURE 8.20
Open-circuit, manually controlled motor drive.

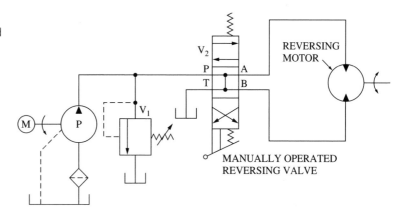

Closed-Circuit Motor Drive

Figure 8.21 shows a one-direction, closed-circuit motor drive. The outlet of the motor is connected directly back to the inlet of the pump. The tank is there to make up any leakage of the system. A pressure filter is added to reduce contamination to the motor. The speed of the motor is adjustable by the variable-delivery pump. The torque is directly proportional to the relief valve setting and to the motor displacement.

Figure 8.22 shows a reversible closed-circuit motor drive. Since the pump and motor will pass flow in either direction, the relief valve and replenishing circuit must function for either condition. When flow from the pump is from line 1 through the motor to line 2, line 1 will be high pressure and line 2 will be low. If the torque requirement of the load causes the pressure of line 1 to exceed the setting of the relief valve, check valves C_3 and C_2 will allow flow to pass over the relief valve at its pressure setting, limiting the output torque of the motor to a safe value.

When the pressure of line 2 is greater than the pressure of line 1 by an amount that exceeds the setting of the relief valve, C_4 and C_1 will allow flow to pass over the relief valve at its pressure setting, thus limiting the reverse torque of the motor to a safe value.

FIGURE 8.21
Closed-circuit motor drive—one direction.

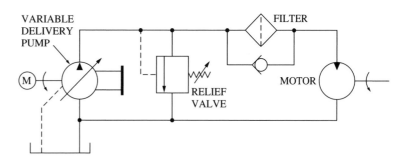

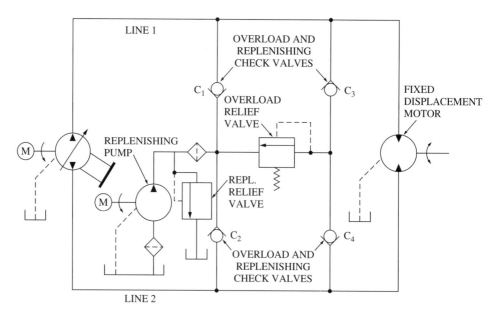

FIGURE 8.22
Reversible closed-circuit motor drive.

Those of you who are familiar with electronics may recognize the check valve circuit of Figure 8.22 to be equivalent to the four diodes used in a bridge rectifier circuit, where line 1 and line 2 would be the ac input and where the relief valve would be the dc load.

The replenishing pump will make up for any leakage in the system. If either line 1 or line 2 needs replenishing, check valve C_1 or C_2 will allow flow from the replenishing pump to the line that needs it.

Characteristics of Closed-Circuit Drives

Closed-circuit drives are designed with various combinations of pumps and motors. Some of these combinations and their resulting characteristics are listed below:

1. *Fixed displacement pump and fixed displacement motor:*

$$\text{Output speed} \times \text{Torque} = \text{Input speed} \times \text{Torque}$$

Select any combination within the horsepower limits of the components.
2. *Variable displacement pump and fixed displacement motor:* The speed is variable by the output of the pump. This drive is sometimes referred to as a *constant torque variable horsepower drive.*
3. *Fixed displacement pump and variable displacement motor:* This drive is sometimes called a *constant horsepower variable torque system.* When the motor is pressure-

compensated, any increase in load torque results in a corresponding proportional decrease in speed.

4. *Variable displacement pump and variable displacement motor:* Provides a very wide speed range with operating characteristics of either a constant torque or constant horsepower drive.

Packaged Drives

Closed-circuit drives are available as integral packages with all controls and valves in a compact housing.

8.6   **SUMMARY**

1. This chapter presents a sampling of commonly used open-loop type hydraulic and electrohydraulic circuits.
2. Dumping through an open-centered or tandem-centered valve reduces the pressure of the pump to near zero, thus unloading the pump and electric drive motor during standby.
3. The speed of the cylinder can be calculated by using the flow to or from either end, provided that the area of that end is known.
4. The identification check for a regenerative cylinder circuit is that the rod end should be connected to pressure while the cap end is alternately connected to pressure (for extension) and tank (for retraction).
5. A system with a 2:1 area ratio regenerative cylinder and a fixed-flow pump has equal extension and retraction speeds.
6. Fixed pressures and noncompensated flow valves provide faster cylinder extension than retraction.
7. The extending speed is 1.414 times the retracting speed when a 2:1 cylinder is cycled from a fixed-pressure source and with a noncompensated bidirectional flow valve connected in the rod end of the cylinder.
8. Remember, to avoid coil burnout, design the electrical circuit so that both solenoids of a solenoid-operated valve cannot be energized at the same time (reference Chapter 4).
9. To move the load down, additional pressure is needed from the pump to overcome the counterbalance valve.
10. Remember that, when there is a choice of cylinder paths, flow will first move the cylinder that requires the least pressure (reference Chapter 5).
11. The horizontal lines of the electrical drawing are numbered on the left side of the drawing. The numbers to the right of the relay coils identify the lines on the drawing where the contacts of that particular relay may be found.
12. *Latching* or *holding* means that the push button may be released and the coil of the control relay will remain actuated through its own contacts.
13. Relay 3CR of Figure 8.19 is known as a sequence relay. It is used to allow the work cylinder to initiate the retraction of the clamp cylinder after the work cylinder has moved forward and tripped 1LS.

14. The four check valves in an overload and replenishing system for a bidirectional closed-circuit motor system are equivalent to the four diodes in an electrical bridge rectifier circuit.

15. Closed-circuit drives are available as integral packages with all controls and valves in a compact housing.

8.7   PROBLEMS AND QUESTIONS

1. How fast will the cylinder of Figure 8.1b extend if the pump delivers 10 gal/min?

2. How fast will the cylinder of Figure 8.1c retract if the pump delivers 10 gal/min?

3. How fast will the cylinder of Figure 8.11 extend if its cap-end area is 10 cm^2 and the pump delivers 10 L/min?

4. How fast will the cylinder of Figure 8.11 retract if its rod-end area is 5 cm^2 and the pump delivers 10 L/min?

5. What would be the minimum storage capacity of the tank of Figure 8.1 if there were five cylinders in the circuit with the same rod and cap areas, but each with a stroke length of 25 in.?

6. What would be the minimum storage capacity of the tank of Figure 8.1 if there were five cylinders, each with a length of 100 cm, a cap-end area of 50 cm^2, and a rod-end area of 25 cm^2?

7. What would be the extending speed of the cylinder of Figure 8.3 if the retracting speed is 250 in./min and all other parameters are the same?

8. What would be the retracting speed of the cylinder of Figure 8.3 if the extending speed is 250 in./min and all other parameters are the same?

9. What would be the extending speed of the cylinder of Figure 8.3 if the retracting speed is 2500 cm/min, while the ΔP retracting across the flow valve = 1,000,000 N/m^2 (1000 kPa) and the ΔP extending is = 2,000,000 N/m^2 (2000 kPa).

10. What would be the retracting speed of the cylinder of Figure 8.3 if the extending speed is 2500 cm/min, while the ΔP retracting across the flow valve = 1,000,000 N/m^2 (1000 kPa) and the ΔP extending is = 2,000,000 N/m^2 (2000 kPa).

11. True or false?: At standby, the pump delivery of Figure 8.1a is being dumped to tank through the tandem-centered, three-position, four-way valve (V$_2$).

12. True or false?: The speed of the cylinder can be calculated by using the flow of either end of the cylinder.

13. True or false?: In the circuits of this chapter, the pump and electric motor are shown on the right-hand side of the drawings.

14. True or false?: In a fixed-flow system the differential cylinder extends faster than it retracts.

15. True or false?: The tandem-centered valve is sometimes used in nontandem circuits.

16. True or false?: In a tandem circuit, the pump is at standby (dumped to tank) when either tandem valve is at center.

17. True or false?: The tank must have the capacity to store all cylinder rods of the system.

18. True or false?: When the cap and rod ends of the cylinder are at the same pressure, the cylinder will extend.

19. True or false?: The result of a 2:1 area ratio regenerative cylinder system with a fixed flow is equal extension and retraction speeds.

20. True or false?: The symbol of a shifted sequence valve would look the same as a standard pressure-reducing valve.

21. True or false?: To prevent coil burnout, design the electrical circuit so that both solenoids of a single valve are energized at the same time.

22. True or false?: Check valve V_3 of Figure 8.6 is used to provide a flow path for P_2 through the unloading valve V_1.

23. True or false?: Check valve V_1 of Figure 8.7 is used to discharge the accumulator.

24. True or false?: The purpose of the CRM contact around the motor start push button of Figure 8.7 is to "seal in" the motor control relay coil.

25. True or false?: The pressure switch of Figure 8.8 both vents and devents the relief valve.

26. True or false?: If you want a simple procedure, all you need to charge an accumulator is a pressure-compensated pump and a check valve.

27. True or false?: In Figure 8.10, the accumulator is discharged automatically when solenoid A of V_2 is energized.

28. True or false?: In Figure 8.11, with counterbalance valve V_3, the load would free fall when V_2 is shifted to the center position.

29. True or false?: The motor of Figure 8.12 will run as a pump during deceleration, because of the load inertia.

30. True or false?: In Figure 8-13, the check valve in the vent line of the V_2 relief valve prevents flow from backing up through the relief valve when solenoid B of V_1 is energized.

31. True or false?: The 50 lb/in.^{2} check valve in Figure 8.14 assures that pilot pressure is always available for V_2, even when it is in the center position.

32. True or false?: When there is a choice of cylinder paths, flow will first move the cylinder that requires the least pressure.

33. True or false?: The small circles to the right of the electrical drawing are the relay coils.

34. True or false?: The vertical parallel lines of the electrical drawing represent the valve solenoids.

35. True or false?: Limit switches in the electrical drawing are always shown in their open position.

36. True or false?: Two small parallel vertical lines with a diagonal line drawn through them is the symbol of a set of NC relay contacts.

37. True or false?: On a sequence diagram, 1CR* is a symbol meaning that "the number one control relay was deenergized at that point."

38. True or false?: The circuit of Figure 8.20 is that of a closed-circuit motor drive.

39. True or false?: The check valve circuit of Figure 8.22 is equivalent to the four diodes used in a bridge rectifier circuit.

40. True or false?: The combination of a variable displacement pump and variable displacement motor provides for a wider speed range than any other combination.

9
Electrohydraulic Servo Systems

Up to now we have considered various hydraulic components, subsystems, and complete systems but all of these have had open-loop control. We now look at "servo" or closed-loop control. With servo the output-controlled parameter is measured with a transducer and fed back to a mixer, which compares the feedback with the command and produces an error signal proportional to the difference. The difference or error signal from the mixer is now used to change the system output until the error is reduced to zero or near zero. *The servo system regulates itself.*

A common example of a servo system is an automatic furnace where a thermostat is used to measure the room temperature and increase or decrease the heat to keep it constant.

To understand electrohydraulic servos, you must first understand the system in which the servo functions. As shown in Figure 7.16 of Chapter 7, the energy needed to power most modern industrial machinery enters the plant in the form of electrical energy. This energy is then converted into other forms for final plant use. Even though the electrical energy is modified in the process of being used to, for instance, mechanical or hydraulic energy, it nevertheless remains the same energy because it obeys the well-established laws of conservation of energy.

9.1   POWER TRANSFER WITHOUT FEEDBACK

Figure 9.1 shows a power transfer system similar to Figure 7.16 except that a speed servo loop has been dashed in and a servo block diagram added. Please disregard these two items for a moment while we examine the need for feedback control.

The electric motor drives a hydraulic pump, which supplies flow and pressure to a hydraulic motor. The motor drives a load through a gear box. Assuming 100% efficiency for all components, the same horsepower that is brought into the electric motor ends up driving the load.

The text of Chapter 9 and most of the figures are drawn from *Electrohydraulic Servo Systems* by James E. Johnson. Copyright © 1973 and 1977 by Penton/IPC; used with permission.

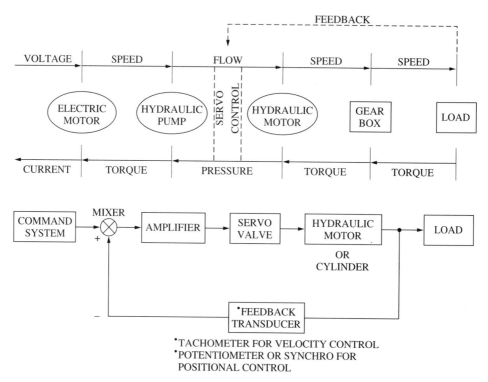

FIGURE 9.1
Typical power transfer system.

Observe again that the parameters across the top of the drawing (voltage, speed, flow, speed, and speed) flow from left to right while those on the bottom (torque, torque, pressure, torque, and current) flow from right to left. This simply means that the input voltage determines the speed of the electric motor, which in turn controls the output flow from the hydraulic pump. The rate of flow from the pump through the hydraulic motor determines hydraulic motor speed, and the speed of the motor (through the gear box) determines the speed of the load.

On the other hand, the torque needed to turn the load reflects back through the gear box and determines the pressure required by the hydraulic motor. The pressure of the hydraulic motor determines the pressure of the pump, which determines the current drawn by the electric motor.

Speed Regulation

To change load speed, you may change any one or combination of parameters shown across the top of Figure 9.1: the ratio of the gear box, flow to the hydraulic motor, pump speed, or the voltage to the dc electric motor. However, changing load speed is not the same as regulating it. There is a marked difference.

To regulate load speed, we must maintain constant speed under all changing load conditions. At first glance this may seem to be a contradiction since it was suggested that the parameters across the top of Figure 9.1 determine load speed. All that was necessary to hold load speed constant (regulate it) was to keep the input voltage constant. This would be true except for the elements shown across the bottom of Figure 9.1.

Torque Fluctuations

Since the object of designing the system is to make it do work, the gear box must generate enough torque to drive the load. The gear box receives its input torque from the hydraulic motor. Should the load increase, the output torque required to move the load will also increase. The increased torque needed at the output shaft of the hydraulic motor requires an increase in hydraulic pressure. An increase in hydraulic pressure means an increase in the torque needed to drive the pump. An increase in torque from the electric motor calls for more current (amperes) from the electric supply.

What does this have to do with speed regulation? Just this: Each of the components of the system, with the exception of the gear box, will tend to slip when its torque or pressure is increased. The hydraulic motor and pump will slip because of the increased internal leakage caused by the increased hydraulic pressure. The increased leakage past the gears, vanes, or pistons in the hydraulic motor and pump will cause the hydraulic motor to slow down. The speed of the electric motor will also "droop" with an increase of torque load. *Because of the slippage of individual components, the system of Figure 9.1 (without feedback) will not maintain a constant speed with a varying load.*

Load Compensation

The design engineer could try to regulate speed by the use of *load compensation*. If the engineer knew in advance that the speed of the system in Figure 9.1 (without feedback) would droop exactly 2% for every 100 lb/in.^{2} pressure buildup at the pump, a load-pressure compensator could be designed for the pump. This compensator would increase the pump outlet flow exactly 2% for each 100 lb/in.^{2} increase of pressure in an effort to keep the speed constant. This is *load compensation*. While it has some advantages, it has serious drawbacks:

1. The viscosity of the fluid changes (the hydraulic fluid becomes "thinner") as its temperature rises. This creates more leakage and the load slows down.
2. The torsional efficiency of the gear box could change with lubrication conditions, and load speed would increase or decrease independently of hydraulic pressure.
3. Electric motor speed may droop with an increase in operating temperature.

So you cannot depend on the 2% speed droop per 100 lb/in.^{2} pressure increase. We must do more.

9.2   POWER TRANSFER WITH FEEDBACK (SERVO)

The best way to regulate load speed is to sense speed and use it to regulate itself. For instance, let's install a transducer (a tachometer generator, for example) at the load to produce a signal voltage proportional to speed. If we feed the signal back to a controller designed to modify one of the appropriate upstream parameters, load speed can be held constant. This feedback-type system is known as a *servo mechanism,* and it has many advantages:

1. Should leakage increase, causing the speed of the hydraulic motor to droop, the tachometer will sense this speed reduction and return the speed back to or near its programmed level.
2. Should the electric line voltage droop, the tachometer will sense a reduction of load speed and make the proper correction.
3. Regardless of the cause, the tachometer will sense any change in speed and make the appropriate correction.
4. Servo drives are easy to program.

Note: The term servo *is derived from "servant," and it denotes a master/servant relationship. The preceding system provides a "command" signal, which tells the system how fast to go. But the system does not stop there. It uses a monitor (feedback tach), which snoops to see if the system really does what it is told to do. If it slows below the command speed, it is speeded up; if it exceeds the command speed, it is slowed down. It is not an open-ended system; it has a closed loop to make sure the output speed is equal to the command.*

Feedback Signals and Control

Servo feedback signals can assume any one of many forms of energy: mechanical, hydraulic, pneumatic, or electrical. Servo signals can also be fed back to any of the components in the chain. *The best method, however, is to feed back an electrical signal to the hydraulic part of the system.*

Low-level electrical feedback signals can be transmitted in the very short time intervals required for reliable and accurate servo control. The components in the hydraulic system respond well to these signals to provide control over a wide range of horsepower. The quick response, low slippage, and inherent stiffness of the electrohydraulic system make it a natural for servo control.

Brief Analysis of Servo

Refer again to Figure 9.1 (with feedback). The tachometer generator is connected from the load to the servo controller.

Observe the closed-loop servo block diagram at the bottom of the figure. A signal that is the difference between the command signal and the feedback signal is fed into the amplifier. This signal is proportional to the system error. The amplifier provides current to the servo valve in proportion to its input signal voltage. The servo valve provides flow to the hydraulic motor in proportion to its input signal current. The motor drives the tachometer, which produces a feedback signal proportional to its speed.

If the load slows down, the tachometer will generate less voltage, thus subtracting less from the command voltage. This will allow the speed to increase to (or near) its set point. If the load speeds up, the tachometer will generate more voltage, thus subtracting more from the command, slowing down the load speed.

To regulate the speed of the load, corrections must be provided for the slippages of all components of the system. This can be done if the sensitivity, or gain, of the servo loop can be made large enough. You might think that all you need do is just crank up the amplifier gain to obtain unlimited accuracy.

Let's draw an analogy. What happens when you excessively crank up the gain (volume) of your speaker amplifier? It becomes unstable and howls. The servo system will do the same thing. It will oscillate and likely do damage to the load or system. The system "loop gain" (the product of all components in the loop) must be kept low enough to prevent instability.

The gain of the servo loop is limited by the lowest resonant frequency of the loop. In essentially all servo systems, whether electrical or hydraulic, the components that limit the frequency response the most are the elements in the power output of the system. For example,

Audio system: speakers.

Electric servos: dc motor and its connected inertia.

Hydraulic servos: hydraulic motor and its connected inertia.

The resonant frequency equation for the hydraulic motor–servo valve system is:

$$\omega h = D_m \times \sqrt{2/VEJ}$$

where

ω_h = hydraulic resonance (1/sec) or radians per sec; f = cycles/sec or hertz = $\omega_h/2\pi$

D_M = motor displacement (in.^{3}/rad)

V = volume of fluid under compression – one-half the volume between the servo valve spool and the motor pistons

E = fluid compressibility (in.^{2}/lb; reciprocal of bulk modules – lb/in.^{2})

J = moment of inertia of the motor plus that reflected by the load (lb-in.-sec^2).

The resonant frequency of the hydraulic valve–motor combination is inherently high for the following reasons:

1. Oil volume under compression can be kept low by close-coupling the servo valve to the motor.
2. Hydraulic motor inertia is low.
3. Oil compressibility is low (4.3×10^{-6} in.^{2}/lb for mobile anti-wear light oil at 130°F).

Low Compressibility of Oil

Low oil compressibility is a key to the response of the servo system. The inertia of any changing load is reflected to the actuator, whether it is hydraulic or electrical. The compressibility of the actuator (hydraulic motor, cylinder, or electric motor) combined with the reflected inertia form a spring-mass resonance.

The lower the compressibility, the higher the resonant frequency. The higher the resonant frequency, the higher the gain of the system can be set without the system becoming unstable. The higher the gain, the more accurately the system can contour or hold a position or velocity.

No variable-speed power transmission exists today (and likely ever will) that can equal the stiffness—and therefore the accuracy—of an electrohydraulic servo system.

9.3   SERVO COMPONENTS

Supply Pumps for Servos

Pressure-compensated pumps are ideally suited for servo power supplies. The servo valve generally requires a constant supply pressure. The compensated pump provides this fixed pressure without the excessive heat and power loss of a fixed-flow pump and relief valve.

Servo Motors

Piston motors are generally better servo actuators than the gear or vane types. The piston-type motor generally has less internal leakage because of the long overlap of the seal area between the piston and bore. In other words, the leakage path is longer and therefore more effective than a shorter path across a vane or gear tooth.

The piston motor generally has a greater breakaway pressure than the vane or gear types. This could cause the motor to ratchet when it is operated in the *open-loop* mode at slow speeds. But ratcheting is no problem in a closed-loop servo system. The excessively high pressure gain of the servo valve at center virtually eliminates this problem, even with an in-line piston motor.

The in-line piston motor has more frictional drag than the bent-axis type. However, this drag does not usually exceed the normal damping required for good servo stability.

Servo Cylinders

Leakage flow and breakaway pressure (required to generate the necessary breakaway force) are two important considerations in selecting servo cylinders. V-type and O-ring seals are generally used for sealing the rods. They do a satisfactory job of preventing external leaks with reasonable friction losses.

Pistons are usually sealed with V-type, T-type, and iron rings. V- and T-type seals require a breakaway pressure about three times greater than iron rings. However, iron rings are likely to allow more leakage. Normally this leakage is moderate and does not exceed the normal damping required for good servo stability.

Servo Transducers

A transducer performs the function of converting a source of energy from one form to another, such as hydraulics to electrical, or mechanical to electrical.

A feedback transducer measures or *samples the output* from a control system, and generates a signal that is fed back into the system and compared with the input (or command) signal. If there is a difference between the command and feedback signals, the system reacts automatically to correct the output until it matches the input command.

The transducer can also be used for instrumentation to measure or record various parameters of a system.

Factors in Selecting a Transducer

Considerations of prime importance in selecting a transducer follow:

1. *Resolution* is the ability to resolve an output signal into a readable value. Transducers with infinite and step-type resolutions are available.
2. *Accuracy* is usually expressed as a percentage of full-scale output. Taken into account are factors such as linearity, temperature effects, hysteresis, and atmospheric conditions.
3. *Repeatability* of the transducer under a given set of ambient operating conditions is often stated as a percent of full scale.

Transducer Types

All transducers, whether they are used as feedback devices or for instrumentation purposes, may be classified in two general categories:

1. *Analog* devices, which produce an output that is *proportional* to the variable being measured.
2. *Digital* devices, which produce an output that may be a count, a series of pulses, or a digitally coded binary number.

Depending on their function in the servo system, transducers may be more specifically classified as:

1. Velocity
2. Positional
3. Pressure
4. Acceleration or
5. Flow.

Figure 9.2 illustrates various types of feedback transducers. Figure 9.2a shows a linear-motion potentiometer used to indicate the position of a cylinder load. Figure 9.2b shows a multi-turn rotary potentiometer, which may be connected to a cylinder load through a rack-and-pinion gear or to a motor load through a step-down gear box.

Figure 9.2c shows a linear variable differential transformer (LVDT). This device operates as a variable transformer. The amplitude of the ac voltage of the secondary is proportional to the position of the core. Starting from null at a position near center, the output voltage has one phase as the core is displaced to the right and a voltage 180° out of phase when the core is displaced to the left.

Figure 9.2d is a pressure transducer, Figure 9.2e shows an accelerometer, and Figure 9.2f shows a torque transducer.

FIGURE 9.2

Types of transducers: (a) Linear-motion potentiometer. (b) Multi-turn potentiometer. (Courtesy of Bourns, Inc.) (c) Linear variable differential transformer. (Courtesy of Schaevitz Engineering.) (d) Pressure transducer produces voltage that is proportional to pressure. (Courtesy of Kulite Semiconductors Products, Inc.) (e) Accelerometer produces voltage that is proportional to deceleration. (Courtesy of Schaevitz Engineering.) (f) Torque transducer has magnetic speed pickup. (Courtesy of Lebow Associates, Inc.)

V_S = SUPPLY VOLTAGE
V_W = WIPER (SIGNAL) VOLTAGE

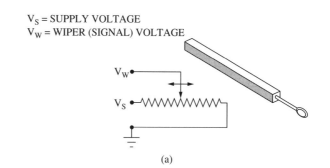

(a)

Bourns, Inc.

(b)

Schaevitz Engineering

(c)

Kulite Semiconductors
Products, Inc.

(d)

Schaevitz Engineering

(e)

Lebow Associates, Inc.

(f)

Servo Valves

The heart of the electrohydraulic servo system is the servo valve, but we should first look at a hydromechanical tracer valve. The follow-type servo valve of Figure 9.3 acts like an electrohydraulic servo valve except that it has a template and stylus to move the valve spool instead of a torque motor.

Both the valve *body* and *spool* of the tracer valve are movable, the spool by means of the template through the stylus and the body by virtue of its connection to the machine slide. The valve spool is spring-loaded to make the stylus follow the template as it is moved up and down.

When the spool of the four-way servo valve is at center, a very small flow is metered from the pressure port, past the A and B cylinder ports to tank. Without the position feedback arm, the pressures at A and B would be equal (about one-half the supply pressure), and the cylinder and load would move to the left.

With the position feedback arm in place, the cylinder will move a very, very small distance to the left until the pressure at B is twice the pressure at A to balance the 2:1 differential cylinder. The load will now follow the template. When the template moves the stylus to the right, the pressure will go up at port A and down at B, moving the load and tracer valve body to the right until the pressures are again at a 2:1 ratio, providing a force balance of the cylinder. When the template moves the stylus to the left, the pressure will increase at port B and decrease at A, moving the load and valve body to the left until the pressures are again at a two-to-one ratio.

The servo or tracer valve, although it is a four-way valve, never completely stops the flow nor fully opens the pressure to the cylinder ports. It is a proportional valve where the load forces are balanced at the center but increase and decrease in either direction proportionally with the spool displacement. *With feedback it is a follow valve.*

What is so important about a follow valve? The load follows the template. Why not let the stylus push the load instead? The answer is that you can push the stylus with your little finger (it has merely to compress the small spring loading the valve spool), but it may require tons of force to move the cylinder load.

FIGURE 9.3

Hydromechanical tracer system illustrating a follow-type servo valve.

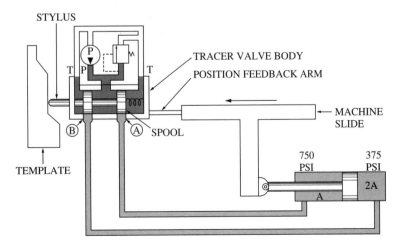

Another example of a hydraulic follow-valve application is in the power steering system of your automobile. The driver of the car turns the steering wheel, which takes the place of the stylus of Figure 9.3 to move the spool inside the power steering valve. The steering cylinder follows the valve spool while "steering" the car. You move the small valve spool while the cylinder uses hydraulic power to "cut" the wheels.

Electrohydraulic Servo Valves

Servo and other proportional valves can be three-way but they are generally four-way valves, or at least have a four-way output stage. The main characteristic of an electrohydraulic servo valve is that its hydraulic output flow amplitude is directly proportional to the amplitude of its electrical dc input current.

Servo valves may be single stage, but when more than a few gallons per minute of flow are needed, two or more stages are generally employed. When more than one stage is used, the pilot (first) stage can take just about any form of design that will produce an output pressure proportional to an electrical input signal.

Figure 9.4a shows a cross-sectional drawing of a single-stage electrohydraulic servo valve. The armature of the dc electric motor is limited, by a built-in torsion spring, to a small arc of travel. The motor is called a *torque* motor because it produces torque,

FIGURE 9.4
(a) Single-stage spool-type servo valve, (b) its symbol, and (c) "quicky" symbol.

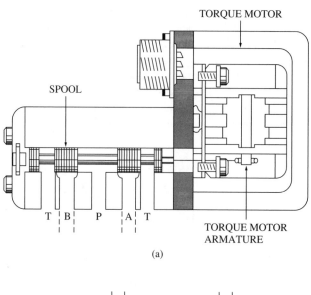

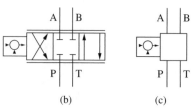

which deflects its torsion spring in a small arc of rotation. A stiff drive wire connects the armature of the torque motor of Figure 9.4a to the end of the valve spool.

The frequency response of a single-stage servo valve is generally higher than that of a multi-stage design. However, the load flow is limited by the power of the torque motor. On rare occasions, very high-powered voice-coil-type force motors have been used to drive the spools of single-stage valves used on hydraulic shakers to vibrate missile nose cones up to 1000 Hz (or 1000 cycles per second).

There is some controversy over which type of first stage to use in a servo valve and over whether to use a first stage at all. Jet-pipe and flapper-nozzle first stages are in common use and do the job nicely for many applications, including many automatic machine tools.

You should obtain technical data from the manufacturers of the servo valves with which you are concerned. Be cautious, however, of the published data—you should understand how it was obtained.

Although two-stage valves have advantages, their big disadvantage is the transport lag of the hydraulic shifting of the second stage. In the interest of developing a small servo valve with a low-powered electric torque motor, some designers have released valves with flow-limited first stages, which means there is just not enough flow from the first stage to move the last stage far enough quickly.

The open-loop response data for these valves may look good at low amplitudes, but ask them to respond fast at high amplitudes and they may fall short of the mark. As stated earlier, if you want a system to test-vibrate the heavy nose cone of a missile at high frequencies, you should select a big single-stage valve with a high-powered voice-coil-type electric force motor to move the single valve spool.

Figure 9.4b shows the accepted symbol for an electrohydraulic servo valve. It depicts a two-stage valve where the first stage is operated electromechanically by the torque motor and the second stage hydraulically by the first stage. The spool is shown as normally closed and the dashed lines along the sides of the spool indicate that the output flow is proportional to the input. After all of this is shown once, it is generally more convenient to just show a servo valve with an electrical input, from an amplifier, feeding a square box with the ports labeled with the usual P, T, A, and B designations (see Figure 9.4c).

Steady-State Valve Characteristics

Ideally, a servo valve produces zero output flow at zero current. In practice, however, this ideal null condition is seldom obtained. The null shift may be due to changes in temperature, supply pressure, or load forces. It is expressed in terms of *null bias* or current changes required to restore zero output flow. In open-loop systems, with normal gain, the null shift of the servo valve is so great that the load drifts and wanders from its assigned position. Because of null shift, a servo valve is seldom used to position a load without a closed loop.

The flow curve of Figure 9.5 is obtained by cycling the servo valve through its rated input current range and recording the continuous plot of output flow for one cycle of input current. The curve shows valve flow gain, linearity, hysteresis, and deadband.

FIGURE 9.5
Flow curve shows hysteresis, deadband, and linearity.

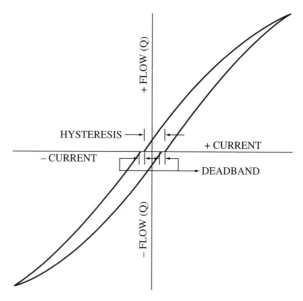

Valve flow gain is the change in output flow per unit change in input current (slope of the curve). This curve is taken at zero or constant output load. Notice that the linearity of the curve changes (the gain becomes less) near the ends of the flow curve.

The hysteresis shows the failure of the flow curve on the way down to follow the curve going up. The deadband is shown as the offset at the center.

Dither

The center condition of the servo valve can be helped by introducing dither current to the torque motor. This means adding a high-frequency electrical signal to the amplifier input. This dither signal must have a combination of a frequency that is high enough and an amplitude low enough that the system load cannot respond to it, but that the valve spool will respond and stay active through center. With dither and a high-loop gain the awful-looking center condition of Figure 9.5 can be tolerated.

Dynamic Valve Characteristics

Valve dynamics can be expressed either in terms of transient (step) response or sinusoidal frequency response. A step response test measures the time required to achieve a desired output, the amount of overshoot, and the settling time. A typical step response curve is shown in Figure 9.6a.

Frequency response tests are a measure of the no-load sinusoidal frequency response characteristics of the valve. Figure 9.6b shows amplitude in decibels (dB) and phase lag in degrees versus frequency in Hertz of a typical servo valve.

*Note: A **dB** is 20 times the logarithm of the signal gain. The gain is doubled for every 6 dB. A **degree** is the measure of a sine wave. There are 360 degrees in one sine wave or cycle. **Hertz** is another name for cycle per second (cps).*

FIGURE 9.6
Dynamic valve characteristics.
(a) Typical curve for the step
response of a servo valve. (b)
No-load sinusoidal frequency
response characteristics.

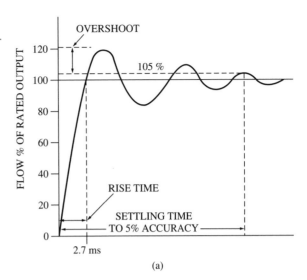

(a)

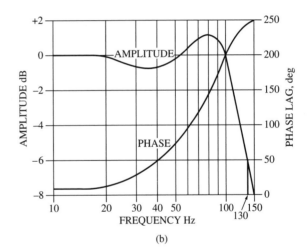

(b)

Effects of Valve Location in a System

Figure 9.7 shows servo valve locations in four different systems. In the system of Figure 9.7a, the valve is close-coupled to the hydraulic motor. This makes it the most accurate and responsive system of the group. It has the least hydraulic fluid under compression in the servo loop, which gives it the highest frequency response of the four systems. Also, as stated before, the higher the resonant frequency of the system, the higher the accuracy and the response. Incidentally, the hydraulic fluid between the pump and valve is not in the servo loop, and therefore does not influence the resonance.

The system of Figure 9.7b is second best. Here, the valve is close-coupled to the cylinder, but there is more fluid under compression than with the close-coupled valve–motor combination.

FIGURE 9.7
Four systems showing different
valve locations.

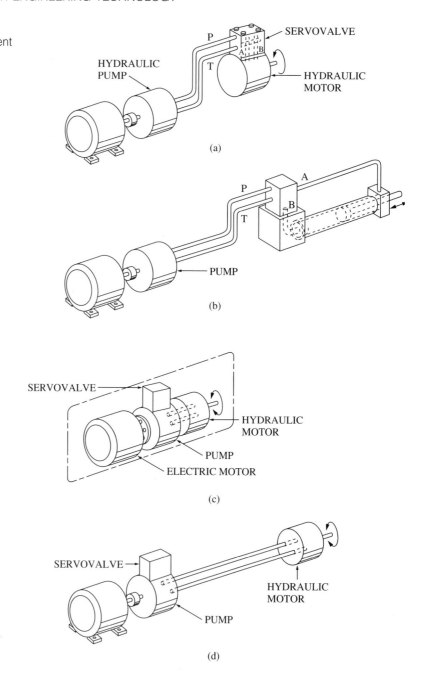

(a)

(b)

(c)

(d)

In the system of Figure 9.7c, the valve is mounted on the pump as part of a yoke-positioning servo system. The hydraulic motor is close-coupled to the pump, minimizing the control fluid under compression. While this arrangement provides a neat, compact, and efficient variable-speed drive, it is not as accurate and responsive as systems A or B (see Table 9.1).

TABLE 9.1
Servo Systems Compared

Type of System (see Fig. 9.7)	Accuracy (in.)	Frequency Response (Hz)	Efficiency
A Valve–motor	0.00001 to 0.002	50 to 150	Good*
B Valve–Cylinder	0.0005 to 0.005	30 to 50	Good*
C Servo pump (packaged)	0.001 to 0.020	10 to 40	Very good
D Servo pump (split)	0.005 to 0.030	5 to 20	Very good

*Efficiency is better when a pressure-compensated pump is used for the supply pressure.

In the system of Figure 9.7d the hydraulic motor is split away from the pump. This allows more versatility. The motor may be located where it's needed on the machine but, due to the excessive volume of control oil under compression (all the oil in both lines between the pump and motor is under compression), this system is the least accurate and least responsive of all four systems in Figure 9.7.

Evaluation of the Four Valve Locations

Table 9.1 ranks the four systems of Figure 9.7 according to general performance. Positional accuracy and frequency response are approximate values and depend on many variables in a specific application (many of which are covered in this chapter).

In addition to high efficiency, the servo pump offers a third controlled dimension to a servo system, that of acceleration. The servo pump acts like a servo valve while controlling the speed and position of the load. But it also provides acceleration control of the load by adjusting the speed of the cylinder, which strokes the pump yoke.

Two-stage Servo Valves

When a second stage is added to a servo valve, feedback must be added between the two stages in order to make the second spool follow the first. Refer back to the follow-type valve of Figure 9.3. If the machine slide were replaced by the spool of a second stage, we could have a two-stage servo valve. The second spool would follow the first (through a gain linkage if more travel is desired of the second spool).

The advantage of a second stage in a servo valve is to utilize the hydraulic muscle power available to move the main spool against high load and flow forces.

Servo Valve in a Positional System

An electrohydraulic positional system is shown in Figure 9.8, along with electrical and hydraulic schematic diagrams. This system could be used, for example, to control a machine slide or table. The amplifier, which is connected between the wipers of the command and feedback potentiometers, sees zero signal voltage when the feedback potentiometer corresponds to the command potentiometer.

When the command potentiometer is moved toward the positive terminal of the power supply, the resulting voltage signal (error signal) is applied to the amplifier, which

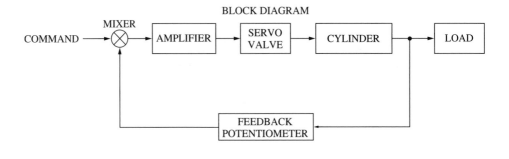

BLOCK DIAGRAM

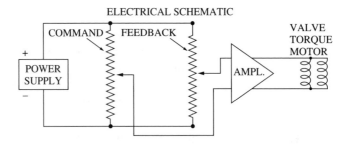

ELECTRICAL SCHEMATIC

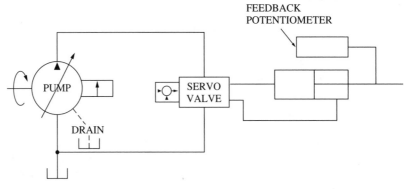

HYDRAULIC SCHEMATIC

FIGURE 9.8
Servo valve in a positional system.

supplies current to the servo valve torque motor. This, in turn, directs oil to the cylinder, causing the cylinder and feedback potentiometer to move and to cancel out the voltage error. The cylinder stops and holds this position until the command voltage is again changed.

Movement of the command potentiometer in a negative direction also results in a movement of the cylinder in the opposite direction until the error voltage is canceled.

This system is essentially an electrohydraulic follower with the feedback potentiometer moving the same electrical distance as the command potentiometer and the load following it. The added advantage of this design over the hydromechanical follower is that voltages from such devices as relays, reed switches, limit switches, numerical control tape readers, and computers can be substituted for the simple command potentiometer voltage.

Servo Valve in a Velocity System

A typical electrohydraulic velocity system is shown in Figure 9.9. The feedback tachometer voltage is connected in opposition to the command voltage, reducing the amplifier output current to the value needed for the desired speed. A typical application of the velocity servo system is to control the speed of a rewind drive in a paper, textile, or metal web processing operation.

BLOCK DIAGRAM

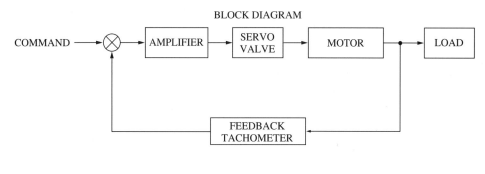

ELECTRICAL SCHEMATIC

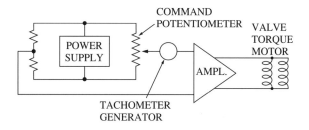

HYDRAULIC SCHEMATIC

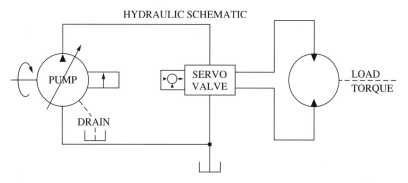

FIGURE 9.9
Servo valve in a velocity system.

When the hydraulic motor attempts to slow down, because of an increase in load, the tachometer voltage will be reduced, canceling less of the command voltage. The higher command voltage will produce more flow to overcome internal leakage of the hydraulic components and slippage of the electric components. This keeps the motor running at or near the desired speed.

When increased speed is desired, the command voltage is increased; the tachometer is driven at a faster speed, producing more voltage to cancel the command.

Servo Amplifiers

The servo amplifier is designed, primarily, to supply proper current to the torque motor of the servo valve and to add gain to the system when needed. The following examples illustrate the advantages, to a system, of using a servo amplifier.

Pot–Pot System Without Amplifier

Figure 9.10 considers a system without an amplifier. The servo valve torque motor is connected directly between the wipers of the command and feedback potentiometers (pots). The torque motor has a current rating of 0.3 A (300 mA) with a 100% overload capacity, or 0.6 A (600 mA). Thus, the maximum voltage that can be supplied to the torque motor is:

$$E = IR \quad \text{or} \quad 0.6 \times 20 = \textbf{12 V}$$

Assuming the stroke of the feedback pot to be 10 in., the voltage gain will be:

$$12 \text{ V}/10 \text{ in.} = \textbf{1.2 V/in.}$$

since:

$$I = E/R \quad \text{or} \quad I = 1.2/40 = \textbf{0.03 A}$$

and the current gain is:

0.03 A/in. or **30 mA/in.**

(Note that total resistance is 40 Ω: 20 Ω for the torque motor, plus 20 Ω for the two pots in parallel.) Assuming a typical 3-mA system deadband, which means that the current must change 3 mA before the load moves, then

$$\text{Positional error} = \frac{\text{in.}}{\frac{30 \text{ mA}}{10}} \times \frac{3 \text{ mA}}{1} = 1/10 \text{ in.} = \textbf{0.1 in.}$$

Thus, the no-amplifier system will repeat a programmed position within plus or minus 0.1 in., which is not very good.

Another drawback to the no-amplifier system, in addition to poor accuracy, is that both potentiometers must carry the full torque motor current and hence must have large power ratings.

FIGURE 9.10
Pot–pot positioning system without amplifier.

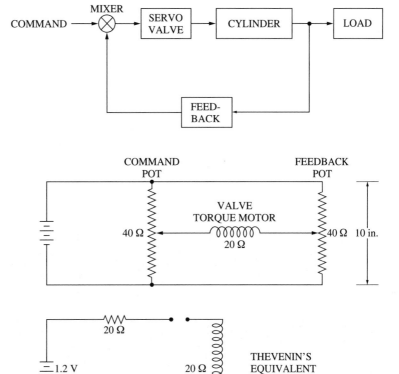

The power rating of the pots would be:

$$P = I^2 \times R$$
$$= 0.6 \times 0.6 \times 40 = \mathbf{14.4\ W}$$

It is hard to find a high-resolution 40-Ω, 15-W feedback pot.

Pot–Pot Positional System with Amplifier

Figure 9.11 shows a pot–pot system with an amplifier. The pots are increased in resistance to 10,000 Ω, which is more suitable for the low input current drawn by the amplifier. The voltage to the pots is also increased to 100 V, which will provide more feedback gain (volts per inch). The wattage drawn by each pot is:

$$\frac{E^2}{R} = \frac{100 \times 100}{10,000} = \frac{10,000}{10,000} = \mathbf{1\ W}$$

The 2-W pots will give us a 100% overrating and are available in many feedback- and command-type units.

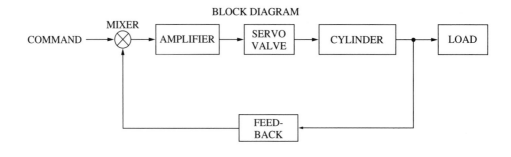

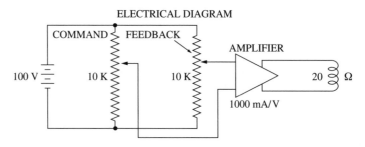

FIGURE 9.11
Pot–pot positioning system with amplifier.

Electrical Gain of System with Amplifier

The travel of the cylinder and feedback pot is 10 in., so the voltage gain of the feedback pot is:

$$100 \text{ V}/10 \text{ in.} = \textbf{10 V/in.}$$

The amplifier gain is adjustable and is set for system stability—say, at 1000 mA/V for this system. The amplifier will saturate at 0.3-V input, so that maximum current sensed by the torque motor is:

$$\frac{1000 \text{ mA}}{\cancel{V}} \times 0.3 \, \cancel{V} = \textbf{300 mA}$$

$$\text{Total current gain} = \text{Feedback gain} \times \text{Amplifier gain}$$

$$= \frac{10 \, \cancel{V}}{\text{in.}} \times \frac{1000 \text{ mA}}{\cancel{V}}$$

$$= \textbf{10,000 mA/in.}$$

Note: The amplifier is designed to saturate (reach its upper current limit) at a value that will not damage the torque motor.

Positional Accuracy with Amplifier

Using the same 3-mA deadband, the error with amplifier would be:

$$= \frac{\text{in.}}{10,000 \; \cancel{\text{mA}}} \times 3 \; \cancel{\text{mA}} = \textbf{0.0003 in.}$$

Thus, a servo system with an amplifier can repeat a programmed position within plus or minus 0.0003 in., which is very good. This represents a substantial improvement over the system without an amplifier. Note, however, that the accuracy is related directly to the gain of the amplifier, which, in this example, was assumed to be 1000 mA/V. As stated earlier in this chapter, the amplifier gain is limited by the lowest resonant frequency of the system. A system with an amplifier gain of 1000 mA/V would be a high-frequency system—one with little mass and a small volume of fluid under compression. More will be said on the subject of "estimating accuracy" later in this chapter.

Operational Amplifiers

Most servo amplifiers are operational or use operational front ends. An operational amplifier can have a very high gain with feedback. Without feedback, the operational amplifier would be unstable because of its excessive open-loop gain.

Unity-Gain Voltage Op Amp

The principle of the operational amplifier is shown in Figure 9.12. Resistors R_i and R_o each equal 1000 Ω. The amplifier obtains its operating power from the supply voltage. Since the operational amplifier has virtually infinite open-loop gain, for all practical purposes, two assumptions can be made:

1. The impedance between the summing junction J and ground is virtually zero.
2. An infinitesimal input signal voltage will produce current in the load.

FIGURE 9.12
Unity-gain operational amplifier.

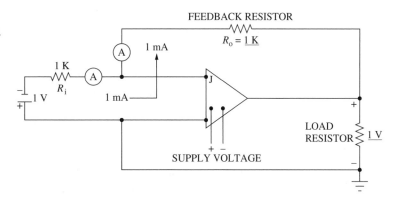

With these assumptions in mind, let us examine the circuit of Figure 9.12. Since J is effectively at ground potential, the 1-V battery forces 1 mA through the 1-kΩ resistor, R_i.

$$I = E/R$$
$$= 1/1000 = 0.001 \text{ A} = \textbf{1 mA}$$

But where is the milli-amp going? Junction J will not accept it. To avoid accepting it, the amplifier drives its output positive, sending the milli-amp around through R_o. Since R_o is 1 kΩ, the amplifier output must become positive 1 V to attract the one milli-amp. *Thus, with a negative voltage applied to R_i, the amplifier produces a positive voltage across the load. This is unity again. It is obtained only when $R_o = R_i$.*

Op Amp with a Voltage Gain of 10

Figure 9.13 is the same as Figure 9.12 except that R_o has been changed from 1 to 10 kΩ. As before, the 1 mA tries to enter junction J. The amplifier rejects the 1 mA and drives the amplifier output to a positive 10 V in order to send the 1 mA around through R_o.

From the examples of Figures 9.12 and 9.13 it is obvious that the gain of the op amp is merely the ratio of R_o/R_i. So to change the gain we need only to change the resistance of R_o.

FIGURE 9.13
Op amp with a gain of 10.

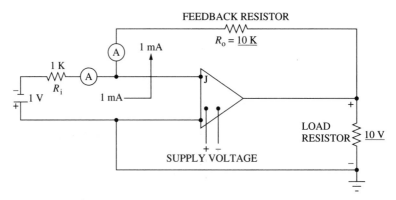

Linear Current Op Amp

Since the servo valve torque or force motor has inductive reactance, the motor current does not quite follow the same changes as the signal voltage. Since the torque of the servo valve is produced by the current drawn by the torque motor, it is imperative that the amplifier output current follow the input signal voltage.

In the amplifier circuit of Figure 9.14, a current-sensing 3-Ω resistor is placed in series with the torque motor load. The voltage from this current-sensing resistor is fed back to summing junction J through a gain control network and feedback resistor R_o. Current through the torque or force motor will now follow the input signal voltage exactly. *The gain of this type of amplifier is expressed as mA/V and is ideally suited to drive torque or force motors.*

FIGURE 9.14
Linear current op amp.

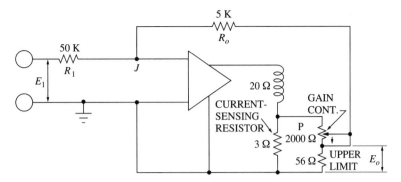

Integrating Current Op Amp

The amplifier circuit of Figure 9.15 is the same as that of Figure 9.14 except that R_o is replaced with two 500-microfarad (µF) electrolytic capacitors connected in series (dc opposing). This arrangement provides a large capacitance (250 µF) in a small package, without any polarization due to electrolytics. Figure 9.15 is an integrating current amplifier and is used in velocity servo control systems. The gain of the amplifier is expressed as mA/sec per volt.

FIGURE 9.15
Integrating current op amp.

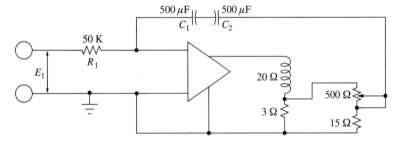

The integrating amplifier will deliver a set current to the torque motor, keeping the speed constant, when the feedback signal voltage is equal to the command voltage. The current will change in a positive direction when the command is greater than the feedback and in a negative direction when the feedback is greater than the command.

When an integrating amplifier is used in a velocity system, velocity errors are changed to acceleration errors. *So, give the amplifier enough time to integrate and it will reduce all velocity errors to zero.*

Figure 9.16 shows a typical application of a linear servo amplifier, and Figure 9.17 shows an application of an integrating servo amplifier. Figure 9.16 is a pot–pot positional servo. Figure 9.17 is a velocity servo with a tachometer generator as feedback.

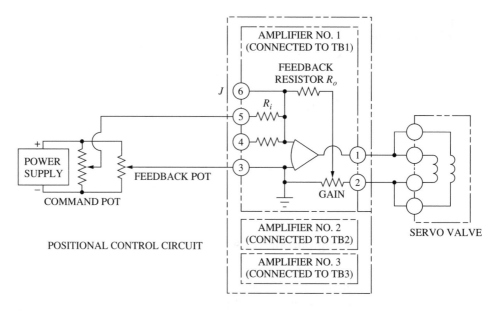

FIGURE 9.16
Typical application of linear servo amplifier. (Courtesy of Vickers Inc.)

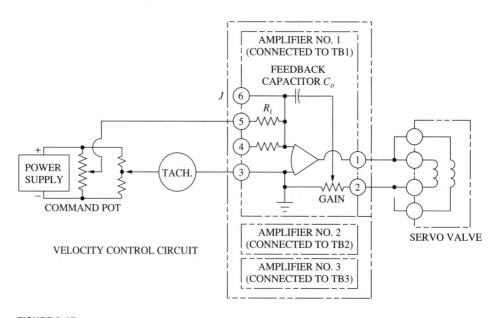

FIGURE 9.17
Typical application of integrating amplifier. (Courtesy of Vickers Inc.)

Servo Programmers

To add versatility to pot–pot servo systems, simulators may be used instead of the standard command pot. Figure 9.18 shows a Kelvin–Varley voltage divider, which simulates a pot with 1000 steps.

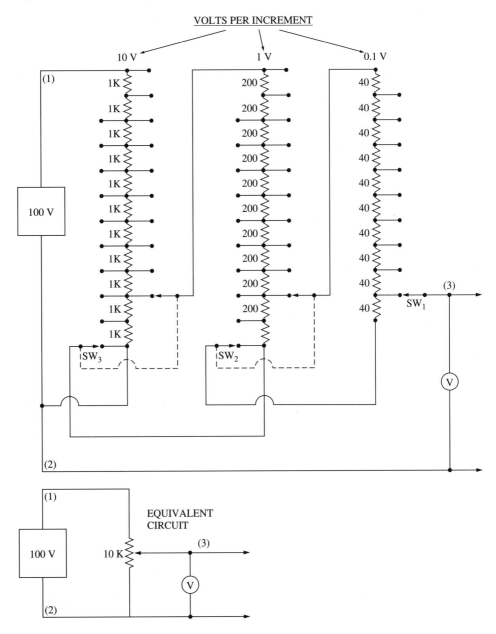

FIGURE 9.18
Kelvin–Varley voltage divider as a pot simulator.

By using three 10-position switches and 32 resistors we obtain 1000 equal steps of voltage changes. The action is like a pot with 1000 position increments.

Let's see how it works. The system has 10 equal resistors of 40 Ω each, in the 0.1 V/step bank. One volt is always applied across the bank regardless of the position of any switch. The voltage output varies from 0 through 0.9 V, in equal increments, as the 10-position (decade) switch SW_1 is moved from bottom to top. Switch SW_2 is a double-pole 10-position switch. As the switch is moved, the two wipers have one contact separating them.

While two adjacent wipers of SW_2 span two 200-Ω resistors at a time (a total of 400 Ω), the actual resistance between the two wipers remains at 200 Ω at all times. The reason for this is that the ten 40-Ω series resistors in the 0.1 V/step bank are also connected across the wipers. This provides a circuit of two branches of resistors, with 400 Ω in each branch, connected in parallel, which equals 200 Ω total. Even though there are 11 resistors in the bank, since the switch spans two at a time, there are only 10 equal resistances in the bank. The same is true for the 10-V bank except that we combine eleven 1-kΩ resistors to equal 10-kΩ total resistance.

There are many ways to simulate a potentiometer to be used in programmers. There are also many other ways to program a servo system. However, space and time limit our further investigation of this topic.

9.4   SERVO ANALYSIS

We have established what a servo is, how it works, and understand some of the parameters that affect the performance of a servo system. We are now ready to analyze a servo drive and estimate its performance.

Analytical Process

To get the most out of the analytical process, think in terms of block diagrams and graphical representations of transfer functions. For example, the block diagram representation for a component is simply

The transfer function (or gain) of a device is the ratio of its output to its input. The following diagram shows how the output is affected by the input:

$$G = \frac{\text{Out}}{\text{In}}$$

Assume an ideal gear box, for example, with a ratio of 1:3; the transfer function is 3. Thus,

$$\frac{\text{Out}}{\text{In}} = 3 \ [(\text{rev/min})/(\text{rev/min})] = \mathbf{3}$$

Similarly, for a pump, the block diagram would be shown as

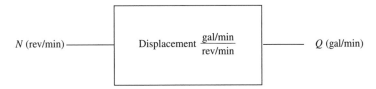

where N is pump speed and Q is the output flow of the pump. Assigning values, if a pump delivers 18 gal/min at 1800 rev/min, the transfer function would be

$$\frac{Q \ (\text{gal/min})}{N \ (\text{rev/min})} = \frac{18}{1800} = \frac{0.01 \ \text{gal/min}}{\text{rev/min}}$$

Then we can multiply input speed by 0.01 and establish output flow.

A servo system automatically corrects its output to correspond to its input or command. The error or difference between output and input is used to correct the output so that the difference is reduced to near zero. When there is information feedback from the output to the input in this manner, we say that the loop is closed or that we have a closed-loop system or servo.

The diagram of Figure 9.19 illustrates a closed-loop servo system in block form. The blocks are a simplified representation of the actual control system components. Block G represents the characteristics of system components between the error signal and output. This is the forward transfer characteristic of the system or forward transfer function. Block H represents the characteristics of the feedback components between the output and input.

FIGURE 9.19
Basic block diagram of a closed-loop circuit.

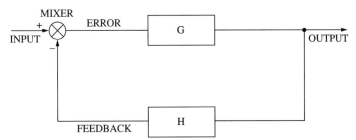

Accuracy and Stability

The primary advantage of a servo control system is that it provides accuracy while maintaining fast cycling time. It also permits continuous adjustment of programming of the output. The degree of accuracy by which the output can follow the input is determined by overall system sensitivity or the magnitude of the open-loop gain. *Prediction of system accuracy is the main objective of this section.*

Note: To obtain open-loop gain we open the loop (by disconnecting the feedback signal at the mixer) and multiply the gains of all components in the loop.

The amount of open-loop gain that can be used in a system is limited. The system will not be controllable if the signal fed back has a greater value and adds directly to the command signal.

Normally in a servo system, output is subtracted from the input. This means that when a sine wave ac voltage is used as a position command, the phase of the feedback signal is shifted 180° and the difference between the command and the feedback, or error, is applied to correct the actual load position error. If the feedback signal is shifted another 180°, the feedback will be in phase with the command signal and the error will increase. When the feedback signal subtracts from the input, it is said that the system has *negative feedback;* if it adds to the input we refer to it as *positive feedback.*

You may ask, "What does this ac have to do with my system? I am using nothing but dc voltage for my command and feedback." The answer is that when there is a disturbance in any spring-mass system (a step change, for example, such as turning on the power), the system reacts sinusoidally at its resonant frequency until the energy of the disturbance is dissipated. During this disturbance, if the feedback is in phase with the input, the system may continue to oscillate at the resonant frequency.

Phase Lag and Amplitude Gain

When using ac analysis, each component contributes phase lag and amplitude gain. If the combined amplitude gain (total open-loop gain) is greater than 1 when the total phase shift is 360°, the system will oscillate.

Phase and amplitude contributions of components generally vary as a function of frequency. We must, therefore, determine both the static and the dynamic characteristics of the components. Figure 9.20 shows a more detailed block representation, including transfer functions (gains) of individual components, for a representative closed-loop electrohydraulic system. The elements in the forward position of the loop (G) include the amplifier, servo valve, and cylinder, while H is the gain of the feedback component.

Open-loop gain is defined as the ratio of feedback signal to input (command) signal with the feedback signal disconnected from the summing point.

$$\text{Open-loop gain} = GH = G_a \times G_{sv} \times G_{cyl} \times H$$

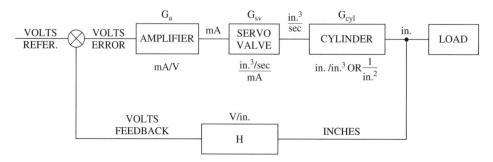

FIGURE 9.20
Detailed block diagram of closed-loop servo.

Another name for GH is velocity constant, K_v. It is found by multiplying all the open-loop gains together:

$$K_v = \frac{\text{mA}}{\text{X}} \times \frac{\text{in.}^3/\text{sec}}{\text{mA}} \times \frac{\text{in.}}{\text{in.}^3} \times \frac{\text{X}}{\text{in.}} = \textbf{1/sec}$$

Prediction of System Performance

To predict the performance of an electrohydraulic servo control system, you must have an understanding of the dynamic, or time-dependent, characteristics of the system components. The *precise* determination of system performance requires that the sinusoidal response of *all* components of the system be determined. The sinusoidal response data consist of amplitude and phase plots versus frequency.

Shortcut to Estimating System Performance

A shortcut to estimating system performance requires that the sinusoidal response data be taken *only* of the component with the *lowest* resonant frequency. Since hydraulic fluid is more compressible than steel, the servo valve–cylinder subsystem has the lowest resonance of the loop (see Figure 9.21).

FIGURE 9.21
Spring-mass resonance of a valve–cylinder subsystem.

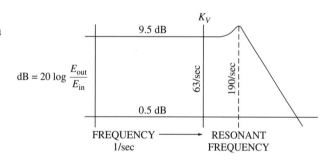

The formula for the resonant frequency of a valve–cylinder system is:

$$\omega_h = \sqrt{2A^2/VEM}$$

where

ω_h = hydraulic resonant frequency (1/sec) or rad/sec

A = annular area of cylinder piston (in.^{2})

V = one-half the total volume of oil in the cylinder plus one-half the volume of oil in both lines (in.^{3})

E = compressibility of oil (5×10^{-6} in.^{2}/lb) (the reciprocal of bulk modules of oil in lb/in.^{2})

M = load mass = $\dfrac{\text{load wt (lb)}}{386 \ (\text{in./sec}^2)}$ = lb-sec^2/in.

Note: To obtain the resonant frequency in Hz, divide the answer of rad/sec or 1/sec by 2π or 6.28. For example,

$$\frac{100 \; \cancel{628 \; rad}}{sec} \times \frac{cycle}{\cancel{6.28 \; rad}} = \frac{100 \; cycles}{sec} \quad or \quad \textbf{100 Hz}$$

Note: To arrive at 386 in./sec^2, the acceleration of a free-falling body, use the physics formula $M = \dfrac{F}{a}$.

Once the resonant frequency is determined for the valve–cylinder, a velocity constant can be assumed for the system. For a typical servo system using a hydraulic cylinder, the velocity constant, K_v, for the system can be assumed to be a value equal to one-third the resonant frequency of the cylinder in radians per second.

Factors Affecting Accuracy

Repeatable accuracy is a function of system deadband (lost motion) of the servo valve plus any other in the system (see Figure 9.5 for valve deadband). The accuracy will be acceptable if the open-loop gain, K_v, can be made large enough. All the deadbands of the system are lumped together (most of it coming from the servo valve) and expressed as milli-amps.

The total error of the system is determined by dividing the total deadband (mA) by the electrical gain (mA/in.) or,

$$\text{Repeatable error } (R_e) \text{ in in.} = \frac{\text{Deadband } (\cancel{mA})}{G_a \; (\cancel{mA}/\cancel{V}) \times H_{fb} \; (\cancel{V}/\text{in.})}$$

Inverting the denominator and multiplying, we have

$$R_e = \cancel{mA} \times \frac{\cancel{V}}{\cancel{mA}} \times \frac{\text{in.}}{\cancel{V}} = \textbf{in.}$$

Note: The terms repeatable error *and* accuracy *are sometimes interchangeable, in usage. When we say that the machine repeats within an error of 0.001 in., we mean that the machine's accuracy is 0.001 in.*

To determine G and H of Figure 9.21, assign some values and analyze the system. The amplifier gain G_a is the last parameter to be established. Since K_v is the product of all open-loop gains, the amplifier gain can be found by dividing K_v by the product of all the other gains:

Note: Remember that the gain or transfer function is equal to the "output" divided by the "input."

$$G_a = \frac{K_v}{G_{sv} \times G_{cyl} \times H_{fb}}$$

where $K_v = \omega/3$.

Step 1: Assume maximum valve current = 300 mA, and maximum flow required = 30 in.^{3}/sec (or 1800 in.^{3}/min); therefore:

$$G_{sv} = \frac{\overset{0.1}{\cancel{30} \text{ in.}^3/\text{sec}}}{\cancel{300} \text{ mA}} = \frac{\textbf{0.1 in.}^3/\textbf{sec}}{\textbf{mA}}$$

Step 2: Assume a maximum cylinder speed of 5 in./sec; therefore:

$$G_{cyl} = \frac{\dfrac{\overset{1}{\cancel{5} \text{ in./sec}}}{\cancel{30} \text{ in.}^3/\text{sec}}}{6 \text{ in.}^2} = \frac{\textbf{1}}{\textbf{6 in.}^2}$$

Step 3: Assume 30 V for 10 in. of travel, therefore:

$$H_{fb} = \frac{\overset{3}{\cancel{30} \text{ V}}}{\cancel{10} \text{ in.}} = \textbf{3 V/in.}$$

Now to set the amplifier gain (mA/V) and estimate the accuracy (in.), we must establish the velocity constant (K_v). To establish K_v, we must go back to the cylinder resonant formula:

$$\omega = \sqrt{2A^2/VEM}$$

Step 4: The rod-end annular area of the cylinder is:

$$\frac{6 \text{ in.}^2}{\dfrac{\cancel{30} \text{ in.}^3/\text{sec}}{\cancel{5} \text{ in./sec}}} = \textbf{6 in.}^2$$

Step 5: Assume 1/2 volume of oil under compression to be:

$$\frac{80 \text{ in.}^3}{2} = \textbf{40 in.}^3$$

Step 6: Assume the weight of the load to be 3860 lb:

$$\text{Mass } (M) = \frac{\overset{10}{\cancel{3860} \text{ lb}}}{\cancel{386} \text{ in./sec}^2}$$
$$= \textbf{10 lb - sec}^2/\textbf{in.}$$

Step 7: Assume the compressibility of oil to be 5×10^{-6} in.^{2}/lb. Then

$$\omega = \sqrt{2A^2/VEM}$$

$$= \sqrt{\frac{2 \times 6 \times 6}{40 \times (5 \times 10^{-6}) \times 10}}$$

$$= \frac{1.414 \times 6}{4.47 \times 10^{-2}}$$

$$= \textbf{189.8/sec} \text{ (cylinder resonant frequency)}$$

and $K_v = 189.8/3$

$$= \textbf{63.2/sec} \text{ (one-third the cylinder resonant frequency)}$$

Thus, the amplifier gain can now be determined:

$$G_a = \frac{K_v}{G_{sv} \times G_{cyl} \times H_{fb}}$$

$$= \frac{63.2}{0.1 \times 0.167 \times 3}$$

$$= \textbf{1261.5 mA/V} \text{ (amplifier gain)}$$

The repeatable error (R_e) in inches may now be determined:

$$R_e = \frac{\text{mA (system deadband)}}{\text{mA/V} \times \text{V/in. (electrical gain)}}$$

$$= \frac{5 \text{ mA}}{1261.5 \text{ mA/V} \times 3 \text{ V/in.}}$$

$$= \textbf{0.00132 in.}$$

(where the unit of inches comes from the bottom of the bottom to the top.) *So, the system will repeat any programmed position within 0.00132 in.*

If a ramp function is programmed into the system, a tracking error will occur proportional to the input current to the servo valve. Assume the input ramp provides a constant current of 10 mA to the valve torque motor. To calculate the error:

$$T_e = \frac{10 \text{ mA}}{1261.5 \text{ mA/V} \times 3 \text{ V/in.}}$$

$$= \textbf{0.00262 in.}$$

So, the cylinder would track 0.00262 in. behind the programmed ramp.

9.5   SERVO SYSTEM APPLICATIONS

This section is aimed at application engineers and designers who use electrohydraulic servo systems; at engineering technicians who test, evaluate, and operate systems; and at maintenance technicians who trim and repair them.

Basic system design concepts are covered in the first part of this section under the Application Notes heading. Later, two systems are described in detail. References are made to other parts of the book when needed for review.

Application Notes

Application Note 1: Cookbook Approach to Accurate Predictions of Pot–Pot Systems

It is possible to predetermine the approximate accuracy of many pot–pot positional systems by using simple rules and tables. The data contained in the tables and the rules given here have been established by extensive testing and from many years of practical applications.

Figure 9.22 shows the electrical schematic of a pot–pot system with a descriptive materials list. The repeatable positional error (P_e) in inches of the system can be calculated from:

$$P_e = \frac{\text{mA (valve deadband and drift)}}{\text{V/in. (feedback gain)} \times \text{mA/V (amplifier gain)}}$$

$$= \frac{5 \ \cancel{\text{mA}}}{5 \ \cancel{V}/\text{in.} \times 1000 \ \cancel{\text{mA/V}}} \text{ (typical)}$$

$$= \textbf{0.001 in.} \text{ (typical)}$$

This repeatable error is predictable for systems with the following limitations:

1. Not more than two-thirds of the supply pressure is used to move the load (the remaining one-third is reserved for the valve pressure drop).

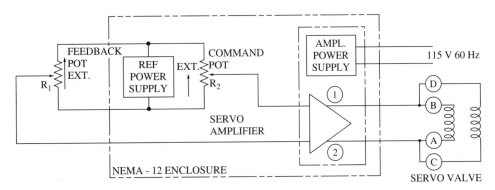

BILL OF MATERIALS

R_1 — COMPOSITION[*] POTENTIOMETER, LINEAR OR ROTARY MOTION WITH
    ANTI-BACKLASH COUPLINGS, 2.5 – 10 K, 2.0 W, 0.5% LINEARITY.
R_2 — COMPOSITION[*] OR WIRE-WOUND, SINGLE OR MULTI-TURN POTENTIOMETER
    2.5 – 10 K, 2.0 W.
REFERENCE POWER SUPPLY (PS) dc POWER, 50 V, 200 mA WITH POTS = 5 K (TYPICAL)
SERVO AMPLIFIER — SUITABLE OUTPUT TO DRIVE SERVOVALVE, INPUT Z = 50 K.
SERVOVALVE — TO MATCH AMPLIFIER AND SUPPLY NEEDED PERFORMANCE.

FIGURE 9.22
Typical pot–pot system.

2. Cylinder speed does not exceed 500 in./min.
3. Load weight does not exceed 1000 lb.
4. Cylinder stroke does not exceed 10 in.
5. Cylinder bore is not smaller than 3¼ in.

Instruction 1: Increase the positional error by 0.001 in. for every 500-lb load increment above 1000 lb.

Instruction 2: Increase the positional error by 0.001 in. for each 10-in. stroke increment beyond the original 10 in.

Instruction 3: Multiply the preceding errors by the factors listed in Table 9.2 for the correct cylinder bore diameter.

TABLE 9.2
Cookbook Error Multipliers

Bore Diameter (in.)	Multiplier
1½	2
2	1.5
2½	1.25
3¼	1
4	0.75
5	0.60
6	0.50

EXAMPLE 9.1: Cookbook Analysis

Assume the pot–pot positional system has a hydraulic supply pressure of 1200 lb/in.^{2}, and must move a 1500-lb load at 300 in./min against a load force of 1000 lb.

Solution: First determine the cylinder bore size needed and the flow required from the servo valve. With an available supply pressure of 1200 lb/in.^{2}, assume that $2/3 \times 1200 = 800$ lb/in.^{2} is available to move the load. Therefore, the minimum effective area of the cylinder is:

$$A \ (\text{in.}^2) = \frac{\text{in.}^2}{800 \ \text{lb}} \times 1000 \ \text{lb}$$

$$= 1.25 \ \text{in.}^2$$

(Reference the information on valve cylinder sizing given in Application Note 2.)

From a standard cylinder catalog, select a 1½-in. bore cylinder which has an annular area of 1.460 in.^{2} and a full bore area of 1.767 in.^{2} (more than the minimum required). The required maximum valve flow rate would be:

$$1.767 \ \text{in.}^2 \times \frac{300 \ \text{in.}}{\text{min}} \times \frac{\text{gal}}{231 \ \text{in.}^3} = 2.295 \ \text{gal/min}$$

A valve rated at 2.5 gal/min at a pressure drop of 400 lb/in.^{2} would be selected because 400 lb/in.^{2} (one-third of P_s) is all that is available for the valve.

The estimated error for this system would be:

0.001 in.	(basic)
+0.001 in.	(additional 500-lb load weight)
	(instruction 1)
+0.001 in.	(additional 10 in. of stroke)
	(instruction 2)
= 0.003 in. × 2	(for 1½-in. bore)
	(instruction 3)
= **0.006 in.**	(maximum repeatable positional error for this system) ■

Application Note 2: Valve-Cylinder Sizing

The hydraulics of a servo system contain three variables:

1. Supply pressure
2. Actuator size
3. Servo valve size.

One of these must be fixed, the other two calculated. A servo application may have a number of solutions depending on which of the three variables is fixed. The following data are needed to specify completely a cylinder and servo valve:

1. Weight of the load W (lb)
2. Stroke length L (in.)
3. Supply pressure P (lb/in.^{2})
4. Load forces F (lb), which are made up of F_c, cutting force, and F_f, frictional force
5. Maximum velocity V_{max} (in./min)
6. Velocity at maximum load V_{load} (in./min).

Cylinder Sizing: The cylinder travel must be slightly more than the required load travel to avoid bottoming before it reaches the farthest target. Usually an extra 0.5-in. of travel is enough. The full-bore area of the cylinder is equal to the needed effective area (plus the cross-sectional area of the rod when the hydraulic pressure must act on an annulus). Dimensionally, the effective area is:

$$\text{Area} = \frac{\text{in.}^2}{\cancel{\text{lb}}} \text{ (Load pressure inverted)} \times \cancel{\text{lb}} \text{ (Load)} = \textbf{in.}^2$$

To guarantee that the servo valve will have an adequate pressure drop to do the job, the cylinder is sized using two-thirds of the supply pressure, with the remaining one-third reserved for the valve. Once the cylinder areas have been determined, the cylinder is selected from a standard catalog. The selection is made such that the cylinder bore selected is equal to or greater than that calculated.

Another consideration in sizing a valve and cylinder is the maximum acceleration permitted by the load. If the maximum acceleration is specified, the product of the supply pressure and the effective area of the piston must be held within acceptable limits:

$$A_{max} = \frac{F_{max}}{M}$$

where A_{max} is the maximum acceleration (in./sec)/sec or in./sec^2, F_{max} is the maximum force available to the load, and M is the mass of the load.

With a step input to the system, the maximum load is equal to the product of the *full* supply pressure and the *effective* area of the cylinder:

$$\frac{lb}{in.^2} \times in.^2 = lb$$

The mass of the load is equal to the load weight divided by 386, or $M = W/386$.

Valve Sizing: Servo valves are rated by flow at a given pressure drop. (Valve manufacturers generally supply flow-versus-pressure drop curves.) The pressure drop available to the valve (P_{sv}) is equal to supply pressure (P_s) minus the load pressure (P_l), or

$$P_{sv} = P_s - P_l$$

Usually, the load pressure varies during the machine cycle, and since the supply pressure is held constant, the pressure available to the valve is reduced as the load is increased.

Care must be taken to make certain that the servo valve has sufficient pressure drop needed to supply its required flow rates. As a rule, two flow-versus-pressure conditions must be considered when selecting a servo valve:

1. *Maximum speed* (usually during rapid traverse) versus valve pressure drop. In this case, the load forces (F_l) are limited to the frictional forces (F_f), or

$$F_l = F_f$$

2. *Speed at maximum load* (usually during "feed") versus valve pressure drop. At this time, the load pressures limiting the valve are frictional forces (F_f) and cutting forces (F_c) or

$$F_l = F_f + F_c$$

■ | **EXAMPLE 9.2: Cylinder Sizing**

Figure 9.23 shows a 1544-lb slide to be positioned over a distance of 10 in. Rapid traverse in both directions is to be approximately 450 in./min. The frictional force is 70 lb. The load must not be accelerated faster than 1500 in./sec^2. The speed at maximum load force is to be at least 180 in./min. The cutting force is 3500 lb, and the supply pressure is 1000 lb/in.^{2}.

Solution: First size the cylinder. The stroke should be greater than 10 in., say, 10.5 in. (Most cylinder manufacturers will supply custom lengths, to fit.)

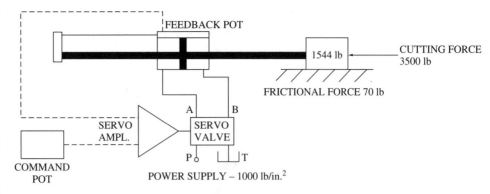

FIGURE 9.23
Loaded valve–cylinder system.

The bore area size is determined by the needed effective cylinder area plus the cross-sectional area of the rod. The effective area is determined by the maximum load force. The *load force* during the *cutting* operation is:

$$F_l = F_f + F_c = 70 + 3500 = \textbf{3570 lb}$$

The *load force* during *rapid traverse* is:

$$F_l = F_f = \textbf{70 lb}$$

The maximum load force is seen during cutting (3570 lb). The supply pressure available to the load is $1000 \times 2/3 = 667$ lb/in.^{2}. The effective area of the cylinder must be at least:

$$\frac{\text{in.}^2}{667 \ \cancel{\text{lb}}} \times 3570 \ \cancel{\text{lb}} = \textbf{5.32 in.}^2$$

Checking a standard cylinder catalog we find that the smallest cylinder that will do the job has a $3\frac{1}{4}$-in. bore with a $1\frac{3}{4}$-in. rod (annular area = 5.891 in.^{2}).

Before going further, check for the *maximum acceleration* of the load:

$$A_{max} = \frac{F_{max}}{M} \qquad\qquad F_{max} = \frac{1000 \ \text{lb}}{\cancel{\text{in.}^2}} \times 5.891 \ \cancel{\text{in.}^2} = \textbf{5891 lb}$$

$$= \frac{5891 \ \cancel{\text{lb}}}{4 \ \cancel{\text{lb}} \text{-} \ \text{sec}^2/\text{in.}}$$

$$= \textbf{1473 in./sec}^2$$

$$M = \frac{W}{a} = \frac{1544 \ \text{lb}}{386 \ \text{in./sec}^2}$$

$$= \textbf{4 lb - sec}^2\textbf{/in.}$$

where *a* is acceleration of a free-falling body.

Therefore, the maximum acceleration of 1500 in./sec^2 is not exceeded. Checking the rod force versus length table (also in a standard catalog), we find that the 1¾-in. rod is more than adequate for the load. The cylinder selected would be 3¼ in. by 10 in. with a 1¾-in. rod.

Now, size the valve from the manufacturer's data. The hydraulic *pressure available* to the valve during *rapid traverse* is:

$$= (1000 \text{ lb/in.}^2) - (70 \text{ lb/5.891 in.}^2)$$
$$= 1000 - 11.9$$
$$= \textbf{988.1 lb/in.}^2$$

Pressure available to the valve during *cutting* is:

$$= (1000 \text{ lb/in.}^2) - (3570 \text{ lb/5.891 in.}^2)$$
$$= 1000 - 607$$
$$= \textbf{393 lb/in.}^2$$

Flow required by the valve during *rapid traverse* is:

$$= \frac{450 \text{ in.}}{\text{min}} \times 5.891 \text{ in.}^2 \times \frac{\text{gal}}{231 \text{ in.}^3}$$
$$= \textbf{11.47 gal/min}$$

Flow required by the valve during *cutting* is:

$$= \frac{180 \text{ in.}}{\text{min}} \times 5.891 \text{ in.}^2 \times \frac{\text{gal}}{231 \text{ in.}^3}$$
$$= \textbf{4.59 gal/min}$$

The valve must deliver 11.47 gal/min at 988.1 lb/in.^{2} and 5.59 gal/min at 393 lb/in.^{2}. Typically, a valve rated at 12.5 gal/min at 1000 lb/in.^{2} would meet both requirements. But, to be absolutely safe, check the flow versus pressure curves supplied by the valve manufacturers. ■

Application Note 3: A Pot–Pot Tracking System

A slave cylinder can be made to track a master by means of a servo system. Figure 9.24 shows a tracking system using a lever-operated, four-way valve to move the master cylinder and a pot–pot servo system to drive the slave. A ratio pot is provided, which reduces the command voltage and thus the distance the slave cylinder must travel to cancel the command. With the ratio pot set at 50%, the slave cylinder will travel only half the distance of the master.

There will be a tracking error as long as the cylinders are in motion. This is due to the fact that there must be an error signal at the amplifier input in order to keep the slave cylinder moving. To calculate the tracking error, the following must be known:

1. Servo valve gain $\dfrac{(\text{in.}^3/\text{sec})}{\text{mA}}$
2. Feedback gain (V/in.)

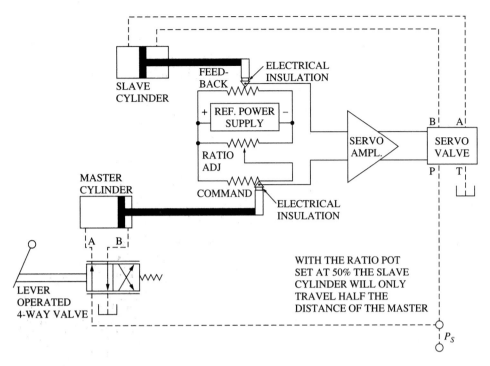

FIGURE 9.24
Pot–pot tracking system.

3. Amplifier gain (mA/V)
4. Cylinder gain (in./in.^{3})
5. Velocity of the slave cylinder (in./sec).

Then, tracking error (in.) is:

$$T_e = \frac{\text{in.}}{\cancel{V}} \times \frac{\cancel{V}}{\cancel{mA}} \times \frac{\cancel{mA}}{\cancel{\text{in.}^3/\text{sec}}} \times \frac{\cancel{\text{in.}^3}}{\cancel{\text{in.}}} \times \frac{\cancel{\text{in.}}}{\cancel{\text{sec}}}$$

$$= \textbf{in.}$$

■ **EXAMPLE 9.3: Tracking Error**

How far behind will a 9.5-in. slave cylinder lag behind its master?

Solution: Assume the effective area of the slave cylinder is 2 in.^{2}. The speed of the slave cylinder is 300 in./min. The 10-in. feedback pot has an excitation of 60 V. The gain of the servo valve is (0.1 in.^{3}/sec)/mA and the gain of the amplifier is set at 800 mA/V. The tracking error is:

$$T_e = \frac{\text{in.}}{6\,\cancel{V}} \times \frac{\cancel{V}}{800\,\cancel{mA}} \times \frac{\cancel{mA}}{0.1\,\cancel{\text{in.}^3}/\text{sec}} \times \frac{2\,\cancel{\text{in.}^3}}{\cancel{\text{in.}}} \times \frac{300\,\cancel{\text{in.}}}{\cancel{\text{min}}} \times \frac{\cancel{\text{min}}}{60\,\cancel{\text{sec}}}$$

$$= \textbf{0.0208 in.}$$

So, the slave cylinder would lag the master by 0.0208 in.
■

Application Note 4: Dual-Gain Positional System

The load of a well-trimmed high-accuracy positional servo system will generally over-shoot its programmed position when given a step command. While this condition is acceptable in many systems, some applications such as machine tools, which must move into the workpiece, cannot tolerate such an overshoot.

The conventional method of reducing overshoot is to lower the amplifier gain. This technique, however, reduces the positional accuracy.

Figure 9.25 shows a circuit that can be added to the operational amplifier of a positional system to remove the overshoot and still provide high positional accuracy.

FIGURE 9.25
Dual-gain system.

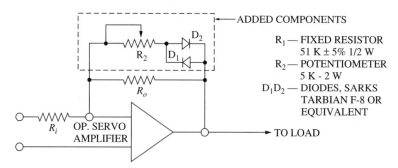

ADDED COMPONENTS

R_1 — FIXED RESISTOR
51 K ± 5% 1/2 W
R_2 — POTENTIOMETER
5 K - 2 W
$D_1 D_2$ — DIODES, SARKS
TARBIAN F-8 OR
EQUIVALENT

The double-gain circuit produces a low-gain condition (and consequently a low deceleration rate) while the load approaches its programmed position. The momentum of the load will have been reduced by this time, thus eliminating the cause of overshooting.

Resistors R_o and R_2 are effectively in parallel, providing low gain to the amplifier, as long as either diode is conducting. As the load approaches its programmed position, the signal voltage to the amplifier approaches zero. When the output signal drops below the threshold of the diodes (0.7 V for silicon), the conducting diode becomes an open circuit dropping R_2 out of the parallel circuit. This raises the gain of the amplifier by increasing its feedback resistance.

Figure 9.26 shows a plot of the output of the amplifier in Figure 9.25 as the load approaches a programmed position. Observe the effect of the circuit on the output volt-age—how it slows the load down to near zero speed (unloading the momentum) with low gain, but brings the gain back up again just before it stops (to ensure positional accuracy).

Resistor R_2 is made variable so that it may be adjusted, during trim, for minimum overshoot.

Application Note 5: Velocity System with Good Steady-State Accuracy

The best steady-state accuracy of a velocity servo system is obtained with an integrating amplifier, as shown in Figure 9.27. An error to the integrating amplifier causes the output current to change at a rate determined by the amplifier gain. This rate continues until the correct output current and, hence, the proper valve flow occurs, reducing the input error to zero. The amplifier then maintains the correct output with zero input.

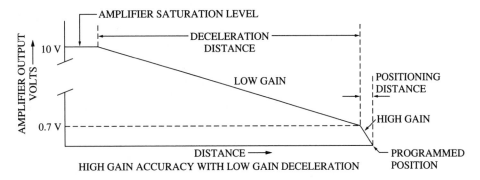

FIGURE 9.26

High-gain accuracy with low-gain deceleration.

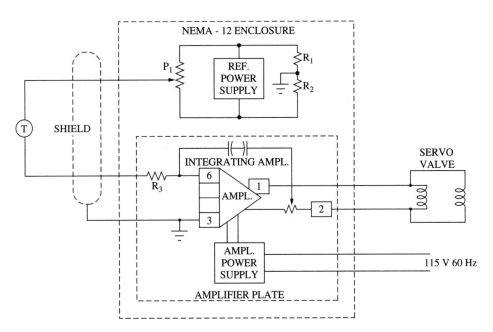

P$_1$ — POTENTIOMETER, COMMAND, ONE-TURN OR MULTITURN 5 TO 10 K, 2 W.
   T — FEEDBACK TACHOMETER, dc.
AMPL — INTEGRATING AMPLIFIER SUITABLE TO DRIVE SERVOVALVE, ADJUSTABLE GAIN
      20 – 1000 mA/V-sec.
REF-PS — REFERENCE POWER SUPPLY REGULATED dc VOLTAGE EQUAL TO TWICE THE
      EXPECTED TACHOMETER VOLTAGE.
R$_1$, R$_2$, — RESISTORS, FIXED, 2 K, 2 W (FOR 120 V MAX. PS).
   R$_3$ — RESISTOR, FIXED, 50 K, 2 W.

FIGURE 9.27

Integrating-amplifier-type velocity system.

If the valve flow changes, due to load or other variations, a momentary speed change will occur until the resulting input error changes the output to the new correct level. For this reason, steady-state velocity error is virtually zero. The only thing that will affect the steady-state accuracy of an integrating-amplifier system is a shift of tach voltage or imperfect regulation of the programmer power supply.

External summing of the command and feedback signals of a velocity system (as shown in Figure 9.27) is recommended wherever possible. This technique electrically unloads the tachometer at null, eliminating most of the tachometer brush noise and voltage changes caused by temperature or brush bounce.

To determine the steady-state accuracy of a velocity servo system with an integrating amplifier, merely add the percent regulation of the reference power supply to the rated percent error of the tachometer. For example,

$$\begin{matrix} \text{Velocity} \\ \text{regulation} \end{matrix} = 0.05\% \ (\% \ \text{power supply regulation}) + 0.05\% \ (\text{tach error}) = \mathbf{0.1\%}$$

The time response of a velocity servo system with integrating amplifier, or the time required to accelerate to a programmed velocity, varies directly with the current (mA) needed by the valve torque motor, inversely with the command voltage, and inversely with the amplifier gain; or:

$$\sec = \frac{\cancel{mA} \ (\text{required valve current})}{\cancel{V} \ (\text{command voltage})} \times \frac{\cancel{V}}{\cancel{mA}/\sec} \ (\text{amplifier gain inverted})$$

A controlled speed range down to 10 rev/min is readily attained, with good steady-state accuracy, using the closed-loop velocity system with an integrating amplifier.

Application Note 6: Velocity System with Good Response

The best response of a velocity servo system is usually obtained with a linear amplifier (Figure 9.28) instead of one of the integrating type. A linear amplifier provides an output current (mA) proportional to its input error voltage. A change in the error voltage causes an instant change in the output current to the servo valve. Response time is therefore not slowed by a rate of output change, unlike in an integrating amplifier.

■ | ### EXAMPLE 9.4: LINEAR-AMPLIFIER VELOCITY SYSTEM

Assume a stable system with the amplifier gain set at 100 mA/V and a feedback gain of 20 V/(1000 rev/min). Assume the rated current of the valve, 300 mA, causes the hydraulic motor, and tach, to turn at 500 rev/min, providing feedback of:

$$\frac{20 \ V}{1000 \ \cancel{\text{rev/min}}} \times 500 \ \cancel{\text{rev/min}} = \mathbf{10 \ V}$$

The amplifier input error would need to be:

$$= \cancel{300}^{3} \ \cancel{mA} \times \frac{V}{\cancel{100} \ \cancel{mA}} = \mathbf{3 \ V}$$

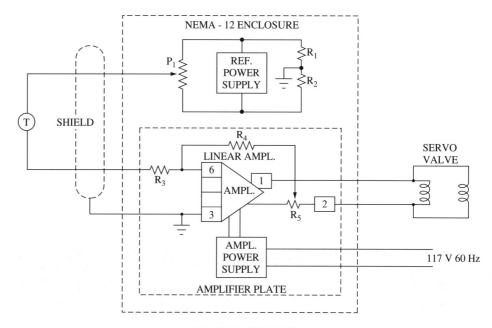

FIGURE 9.28
Linear-amplifier-type velocity system.

The command voltage needed to produce a speed of 500 rev/min is 3 V more than the feedback of 10 V or **13 V** total.

At one-half full speed, the feedback voltage is one-half 10 V or **5 V**. The valve current is one-half 300 mA or **150 mA,** and, the command voltage would be one-half 13 V or **6.5 V.**   ■

■   **EXAMPLE 9.5: Speed Droop**

To calculate the speed droop caused by an increase of motor load, let's first assume an open-loop droop of 5%. If the load speed is 500 rev/min before the increase in load, the speed would droop $500 \times 0.05 = $ **25 rev/min** (open loop).

Feedback will reduce the open-loop droop by the ratio of the feedback signal to the command signal.

The actual droop with feedback will be:

$$25 \text{ rev/min (open-loop droop)} = (25 \text{ rev/min}) - \frac{10 \text{ V (tach)}}{13 \text{ V (command)}} \times 25$$

$$= 25 - 19.23 = \mathbf{5.77 \text{ rev/min}} \qquad \blacksquare$$

■ | **EXAMPLE 9.6: Closed-Loop Accuracy**

Since this system has a linear amplifier, we must consider its drift as well as the valve deadband while calculating the total error. The valve–amplifier error can be estimated at 1% of full valve current (or 3 mA). This would result in an additional error of:

$$\frac{1000 \text{ rev/min}}{20 \text{ V}} \times \frac{\cancel{V}}{100 \text{ mA}} \times 3 \text{ mA} = \mathbf{1.5 \text{ rev/min}}$$

The total closed-loop error would be 5.77 rev/min + 1.5 rev/min = **7.27 rev/min.** This is typical accuracy for a linear-amplifier-type velocity servo system. However, if accuracy is the only criterion for using a servo system, you may want to consider using a pressure-compensated flow control valve instead. ■

9.6   EXAMPLES OF SERVO APPLICATIONS

Servo Level Control of NASA Crawler

Electrohydraulic servo systems are used extensively on aircraft and space vehicles as well as their ground-handling equipment. The giant muscle power of hydraulics controlled by the split-second timing of electronics serves the missile and space industry well.

Figure 9.29 shows a sketch and basic diagram of the leveling control system for the transporter–crawler designed to carry the Apollo/Saturn V rocket (including the launch tower) from the assembly building to launch pad 39 at Kennedy Space Center, Florida. The crawler has been used after Apollo to move the shuttles and their towers from the assembly building prior to takeoff.

The crawler was designed to carry 15,000,000 pounds, and travels on two parallel road beds of crushed rock. A square platform with a diagonal measure of 119 ft is supported by four hydraulic jacks (one at each corner) and by a lower frame mounted on four tracked vehicles. To further understand the enormous size of the vehicle, each cleat of the caterpillar-like tracks weighs approximately one ton.

The four jack cylinders are operated in unison (circuit not shown) to pick up the launch tower and missile from the assembly building and deposit them at the launch site. Once the load has been picked up, it must be kept perpendicular as it is transported over the resilient (gravel) roadway to pad 39. The sketches and basic circuit of Figure 9.29 show the automatic servo level control.

A spherical reservoir half-full of water is placed near the center of gravity of the crawler and higher than the four pressure transducers placed at the four corners of the platform. This placement ensures that the pressure at each transducer is higher than that of the reservoir. The water system is purged to remove any air from the transducer lines. The output of diagonally opposite transducers is shown connected in series-bucking. With

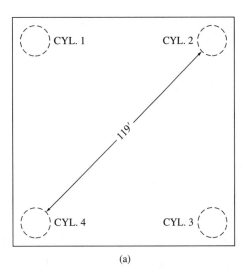

(a)

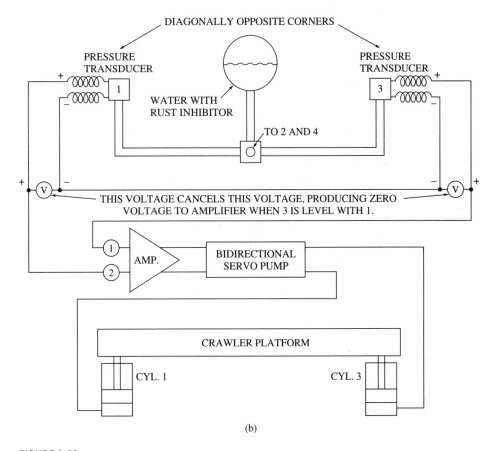

(b)

FIGURE 9.29
Crawler–transporter level control.

this connection, the signal voltage to the servo amplifier is zero when the two transducers (1 and 3) are at the same height (level). If transducer 3 should become higher than 1, the voltage at terminal 1 of the amplifier would become less positive, causing the dual-acting servo pump to pump oil from cylinder 3 to 1, thus leveling the platform. The opposite would be true if cylinder 1 should become higher than 3. A second leveling circuit, just like the one shown for leveling cylinders 1 and 3, is used to level cylinders 2 and 4.

Servo systems are also used in many processing industries such as metals, plastics, paper, meat, and textiles. Two systems used in metals processing are described here as examples. The first is a system for process control of continuous billet castings. The second is a system that provides level-wind control of metal tubing while it is being wound into a neat roll for final annealing and shipment. The two systems cover the very beginning and very end of metal tube manufacture.

Servo Control of Continuous Billet Casting

In the continuous billet casting process (Figure 9.30), molten metal is siphoned from a ladle into a mold where it is cooled and drawn as a continuous billet. The mold casts the billet as a continuous tube.

Pinch rolls, powered by hydraulic motors, pull the billet from the mold where it is bent 90° and fed into straightening rolls. The straightened billet is then cut to length by a flying saw.

An electrohydraulic velocity servo system controls the flow through the pinch-roll drive motors. A speed control pot supplies the command voltage to the servo amplifier. The amplifier drives the servo valve, which supplies oil flow to the pinch-roll drive motors and the tachometer motor, all in series. When the drive reaches command speed, the pinch-roll tachometer stops the acceleration and maintains nearly constant speed.

The voltage from the pinch-roll feedback tachometer is also fed through a dancer pot, and acts as the command voltage to the straightening-roll servo system. The straightening-roll system also has a separate series-connected hydraulic motor to drive its feedback tachometer.

The dancer pot senses the radius of the bend in the billet and accelerates or decelerates the straightening roll drive to maintain the proper bending radius.

In other words, one speed pot controls both drives. The speed of the straightening-roll drive follows the speed of the pinch-roll drive because the pinch-roll feedback tachometer acts as the command to the straightening-roll drive. The dancer pot assures absolute tracking.

Separate hydraulic motors are used to power the tachometer generators for two reasons: (1) The system may be operated without a billet for setup purposes and (2) *more importantly,* the system will operate with higher gain and improved stability because the mass of the load has been taken out of the servo loop. To put it another way, the drives are not really *velocity* servo systems but rather *flow* servo systems. The tachometers sense the flow through the pinch and straightening drive motors but not their true speed.

The drive motors of both systems are connected in series and are far apart. This puts an enormous volume of oil under compression between the ports of the servo valve. If the tachometers were placed on the billet, providing true velocity servo, the spring-mass resonance would be so low that effective control would be impossible. However,

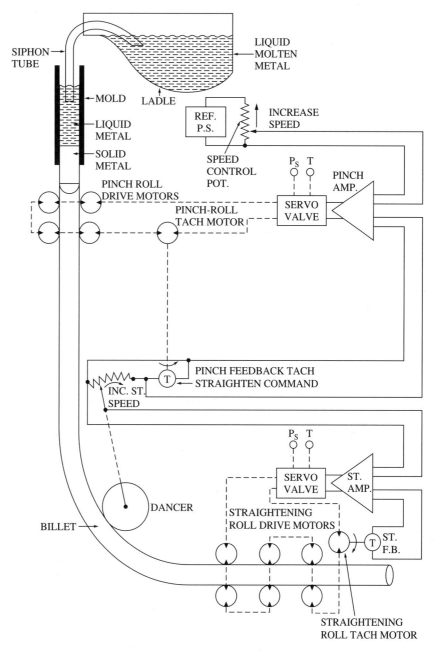

FIGURE 9.30
Servo control of continuous billet casting.

with separate tach motors, the total inertia is not reflected back into the servo loop. As the load mass is accelerated, leakage increases, dampening the effect that the reflected mass has on the fluid. Higher gain, therefore, can be set for the amplifier, providing more accurate control of hydraulic flow.

When the temperature of the oil increases, the motor speed will droop due to the increased internal leakage of thinner oil; yet, the flow through the tach motor will remain about the same. The stability and tracking of the two drives are much more important than the absolute speed of the system. Any speed droop due to oil temperature increases can be corrected by the operator as long as the system is stable and tracking.

Liquid Level Control

Liquid level in the mold is determined by the balance between the molten liquid flow into the mold and the casting withdrawal from the mold. Automatic control of this liquid level is important for a number of reasons:

1. Startup reliability.
2. Operating reliability.
3. Metal grain size control.
4. Freezing near bottom of the mold.
5. Overreaction of operator.

Metal flow into the mold is accomplished with a tilting ladle and siphon arrangement. The flow rate is established by varying the head of liquid metal to the siphon tube by tilting the pouring ladle. Figure 9.31 shows a block diagram for the liquid level control. A tilt command voltage is fed into the servo amplifier, which drives the servo valve that ports oil to the hydraulic cylinder. The cylinder tilts the ladle, moving the feedback pot wiper and canceling the command voltage. This completes the *minor* loop, providing a position follower. The ladle tilt position follows the manual-tilt control pot.

The major loop consists of a preamplifier, which receives its command signal from a liquid level setting pot and feeds the minor loop. The output from the minor loop provides "inches of ladle tilt" as a function of "input volts." As the ladle tilts, liquid metal flow increases into the mold, raising the liquid level and increasing the signal voltage from the liquid level transducer, thus canceling the liquid level command voltage stopping the tilt. We now have another position follower. The liquid level in the mold follows the liquid level command fed into the preamplifier.

Servo Control of Automatic Level-Wind System

As the takeup reel of Figure 9.32 rotates, the pulse generator, A, supplies 120 pulses per revolution to the frequency divider, B. The frequency divider supplies 20 pulses per revolution (assuming 0.5-in. tubing) through the forward gate of preset counter C and through driver amplifier D to the forward coil of stepping motor E. The stepping motor rotates the command synchro, F, 1/10 revolution per revolution of the takeup reel. Since

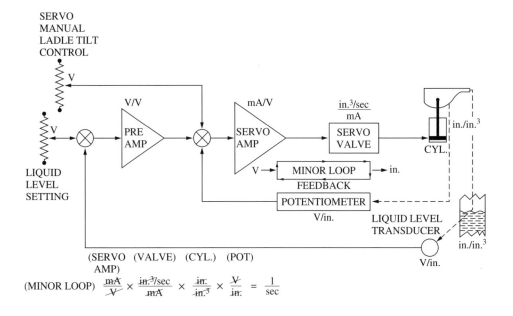

FIGURE 9.31
Block diagram for liquid level control.

the stepping motor requires 200 pulses per revolution, the 0.5-in. tubing with its 20 pulses per reel rotation will rotate the command synchro:

$$\frac{\text{rev}}{\cancel{200}\,\cancel{\text{p}}} \times \cancel{20}\,\cancel{\text{p}} = \textbf{1/10 rev}$$

As the command synchro is rotated, an error voltage is generated that is a sine function of the angular difference between the command synchro and feedback synchro G. This sine wave signal is fed to servo amplifier J through demodulator H, where it is converted to a dc signal suitable for the amplifier. The amplifier feeds the servo valve, which drives the traversing cylinder, moving the rack 0.5 in., and rotating the feedback synchro 1/10 revolution, thus canceling the synchro error. The rack rotates the pinion one revolution per 5 in. of cylinder travel or 1/10 revolution per 0.5 in. of travel.

Without the digital-to-analog (D/A) converter, M, the feedback synchro would lag behind the command synchro by an amount proportional to the speed of the cylinder. The reason for this velocity lag is that the cylinder requires hydraulic flow as long as it is moving. For the servo valve to supply this flow, it must have current from the amplifier, which must have an error voltage from the synchros.

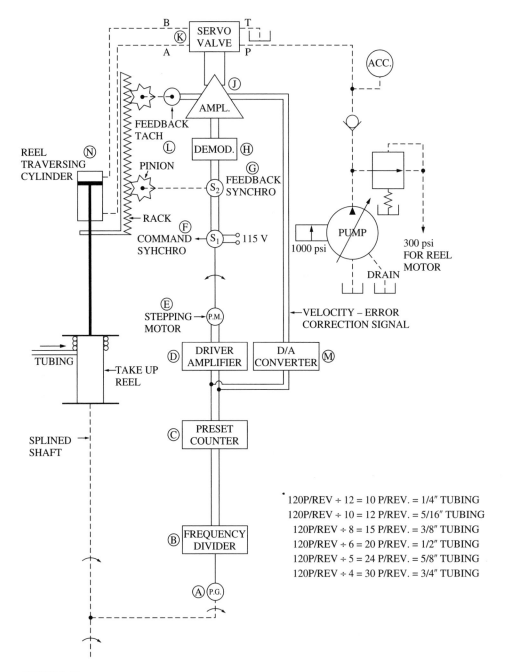

FIGURE 9.32
Automatic level-wind system.

The D/A converter supplies the amplifier with a signal voltage proportional to the pulse rate and therefore to the cylinder velocity. The signal voltage from the D/A converter is trimmed to compensate for the velocity lag to bring the synchros into phase at all speeds.

Another way of looking at the performance of the system is that the pulse rate, through the D/A converter, programs the velocity of the traversing cylinder. The synchros provide positional tracking control so that the takeup reel is traversed a distance exactly equal to the diameter of the tubing during each wrap.

When the counter reaches a preset number, indicating that the wrapped tube layer has reached the end of the reel, the pulses are removed from the D/A converter and from the driver amplifier, stopping the cylinder. The cylinder remains in the end position for 20 pulses, for the 0.5-in. tubing, allowing one complete wrap to be placed over the end wrap of the previous layer before reversing the cylinder. At the second preset of the counter (first preset + 20 counts) the pulses are again applied to the driver ampli-

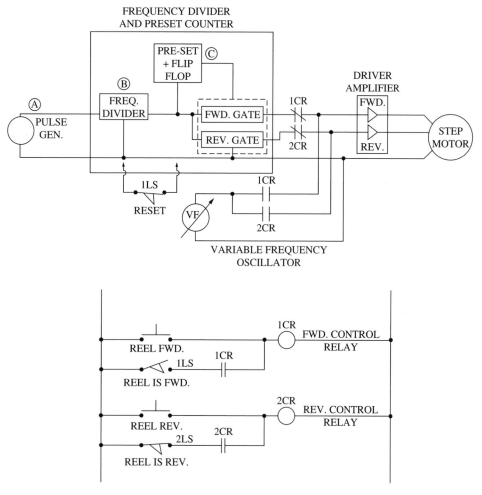

FIGURE 9.33
Stepping motor drive circuit.

fier, but to a second input, which causes the stepping motor to step in the opposite direction.

The system repeats this cycle until the reel is full. At this time the rotation of the takeup reel and pulse generator stop, bringing the traversing system to rest.

The traversing system is now operated manually to remove the coil. This is done by feeding pulses from a variable-frequency oscillator (VFO) through the driver amplifier to the stepping motor (Figure 9.33).

When the wound coil is removed and a new one is to be started, the "Reel forward" button is pushed. This energizes 1CR, applying the VFO signal to the forward input of the driver amplifier. Relay 1CR is sealed through the "Reel is forward" limit switch 1LS. The takeup reel is traversed to its starting position where it trips 1LS, dropping 1CR and resetting the counter to zero. The next reel is then wrapped automatically when the reel starts turning.

9.7   SERVO TRIM

Trim procedures are usually covered extensively by manufacturers of servo components and systems. However, their service brochures generally apply to their own specific equipment. They usually have their own names for component parts and adjustments, which can make it somewhat difficult to carry over the knowledge learned from one manufacturer to the product of another.

This section covers system trim in a general way; it is specific only when necessary to explain the concept. It is not the intention of this section to replace manufacturers' instructions but to supplement them.

Trimming the Servo System

Three rules should be observed when trimming a servo system:

1. Know your servo system.
2. Know the machine the servo is designed to control.
3. Proceed with caution.

Trimming a servo system can be like breaking a horse. The system has a lot of responsive energy, and if released without proper control, there could be trouble. *For the initial startup, the hydraulic control pressure should be reduced to a level that will barely move the load.*

Proper Phasing

The first step in trimming a system is to check for hydraulic leaks and proper phasing. After the system is properly installed, reduce the pressure setting and jog the system (momentarily start the system and then, quickly, turn it off). Check for leaks! If the oil is not spilling onto the floor, restart the system at low pressure and check for proper phasing.

Proper phase for a typical system is shown in Figure 9.34. When the command pot wiper is moved up, a positive voltage is produced at the input of the operational amplifier, A. A negative voltage is applied to the servo valve, B. The servo valve supplies (reduced) pressure to the cap end of the cylinder, C. The cylinder should react by moving the load and wiper of the feedback pot, D, toward the positive terminal of the reference supply, thus rebalancing the amplifier. *If the cylinder does not move, slightly raise the hydraulic pressure (toward normal) until it does move.* If the cylinder moves the load to its programmed position and stops, the system is either properly phased or two (or all four) of the following sets of connections are reversed.

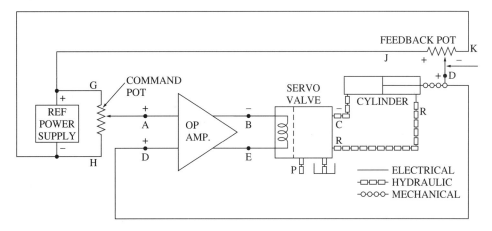

FIGURE 9.34
Proper phasing in a servo system.

1. Amplifier input terminals A and D.
2. Servo valve input terminals B and E.
3. Servo valve output ports C and R.
4. Feedback pot excitation terminals J and K.

If the cylinder goes hard over to either end and stops, then one or three of the above sets of connections are reversed.

The reason that the cylinder would have gone hard over if one or three of the above sets of terminals were reversed is that these reversals would have caused the feedback to add to the error instead of reducing it (positive feedback instead of negative).

If the system is found to be stable but the cylinder retracts when it is programmed to extend, one of two solutions would solve the problem:

1. Reverse the excitation leads (outer leads) of the command pot G and H.
2. Reverse the outer leads of feedback pot J and K along with one other set of connections listed above.

Setting the Gain

If the gain of the system is too high, the system will become unstable and oscillate at its natural frequency—like a public address system with too much volume. The best way to adjust system gain is to start with the amplifier gain control at minimum (Figure 9.35). Increase hydraulic pressure to its rated value. Then increase the amplifier gain slowly until the system becomes erratic or starts to oscillate. Now lower the amplifier gain about halfway and observe that there is no tendency for erratic motion or instability.

If it is difficult or impossible to set the gain because of an excessively imbalanced amplifier or valve, set the gain to minimum and proceed to null as described in a later section on Valve and Amplifier Null Adjustments. After this preliminary null, go back to the gain setting and proceed as before.

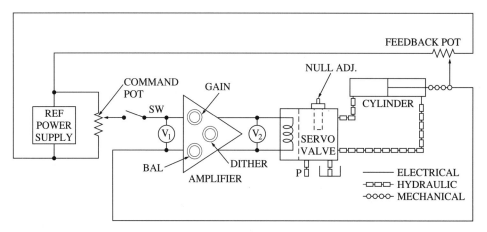

FIGURE 9.35
Trimming the servo system.

Dither Adjustment

The dither adjustment (Figure 9.35) allows the system to be trimmed so that the valve spool moves at the high dither frequency but at such a low amplitude that the load does not respond to it. This reduces the valve deadband with no detrimental effect on the load. To adjust dither, observe the following:

1. With a small load weight (one that can be moved by the dither amplitude of less than 10% of amplifier saturation), increase the dither until the load responds. Then lower the dither until there is no detectable response.
2. With a large load weight (one that will not respond to 10% dither), cycle the load, approaching the same programmed position from both directions, and observe the positional error. Increase the dither until the error reaches a minimum. Do not increase the dither further. Do not add more dither than the valve needs, because the excess will only cause unnecessary valve wear.

Valve and Amplifier Null Adjustments

Now that the gain and dither have been set, we can null the valve and amplifier (refer again to Figure 9.35).

1. With the hydraulics turned off, the amplifier turned on, and the switch (SW) turned off, adjust the amplifier balance control until voltmeter V_2 reads zero. Note that you should be able to read virtually zero on the 3-V scale of a voltmeter for torque motors rated at 5 to 15 V.
2. With everything turned on and the system regulating a position, turn the servo valve null adjustment (mechanical) until voltmeter V_2 again reads zero.

This means that with a single-rod cylinder the servo valve has been mechanically offset, slightly, to provide more hydraulic pressure on the rod end of the cylinder. This allows the electrical circuit to operate at zero null. The advantage of this arrangement is evident when we consider that electrical drift of a device is a percentage of the output voltage or current of the device. If output is zero, then drift is zero.

9.8 SERVO MAINTENANCE

Maintenance of electrohydraulic servo systems, including computers and microprocessor controllers, is one of the most challenging technologies of the computer age. Most of the machines controlled by servos are expensive and downtime is costly. This puts a premium on efficient and reliable maintenance. Obviously it would be impossible to do justice to the entire field of servo maintenance in one section of one chapter of this book. Here we will simply attempt to provide some basic service tips and service information collected through years of experience in the field.

Four Important Service Tips

Tip 1: You can no longer be just an electronics technician or just a hydraulics technician and survive in the servo world. You must have a working knowledge of the inputs and outputs of both electronics and hydraulics. You must be a systems person. You must know about the brains that program the servo and also the muscle that does the work.

The same underlying principle applies to mechanics, hydraulics, and electronics. All three fields are concerned with a force moving something against an opposition and, in all cases, Result = Effort/Opposition. This is the basis of physics. Hydraulics obeys the law, and electronics engineers know it as Ohm's law. The following table shows the relationship:

	Result	**Effort**	**Opposition**
Mechanics	Velocity (in.min)	Force	Resistance
	Speed (rev/min)	Torque	Friction
Hydraulics	Flow	Pressure	Restriction
Electronics	Current	Voltage	Resistance
	(*I*)	(*E*)	(*R*)

Tip 2: Spend an adequate amount of time diagnosing the trouble. Don't jump to conclusions. More time is wasted and more systems damaged by snap judgments than by lack of knowledge. Use the rule of thumb that there must be two or more nonconflicting indications that the trouble is in a certain subsystem or component before you "tear into" that subsystem or component.

For example, do not disassemble the servo valve just because you are not getting hydraulic flow to the cylinder. You should first check hydraulic pressure to the valve, the voltage to the torque motor, and the continuity of the involved electrical and hydraulic circuits.

Tip 3: Obtain the latest and most complete manufacturers' information. Too much time is spent guessing at specifications when a little regularly done bookkeeping or a phone call would eliminate all doubts.

Tip 4: Do not lose your cool. Keep a record of what you have done. If possible, work with someone as a team. You will help keep each other on track.

Ten Sources of Contaminant Troubles

Now we consider some technical service information. Perhaps the one thing that contributes most to long, trouble-free servo life is a good procedure for flushing the hydraulics system.

The sources of contaminants that can enter a hydraulic system include:

1. Grit or dust in the atmosphere
2. Core sand dislodged from castings
3. Metal particles from pumps, valves, and fittings
4. Particles from packings
5. Textile lint
6. Cutting oils
7. Water
8. Pipe joint compound
9. Paint
10. Products of oil decomposition.

These particles are often so fine that they do not settle out, but remain in suspension, eventually wedging themselves in the close diametral clearances of servo valves. It is essential that these fine clearances remain open to achieve the finite control, repeatability, and accuracy necessary in a reliable servo system.

Preliminary Flushing

Preliminary flushing is strongly recommended to ensure the removal of any contaminants "built into" the system, including the so-called "clean" oil, which can be very contaminated from the barrel. Without flushing, metallic chips, thread compounds, core sand, etc., can cause serious damage—if not complete failure of the system—within the first two hours of operation.

1. Clean the reservoir with a dry-cleaning solvent. Use filtered air to remove solvent residue. (Federal Specification P-S-661 is a satisfactory solution.)
2. Clean all lines that are used in the system. In some cases pickling of lines and fittings is necessary.
3. Install full-flow filters with ratings not larger than 10 μm to the pressure and main supply lines of the servo valve.
4. Install a flushing plate to the manifold in lieu of the servo valve.
5. Check to see that all lines are adequately sized and attached to the proper connections.

Note: The servo valve flushing plate permits the fluid to flow from the pressure line to the manifold and directly back to the reservoir. This allows the fluid to recirculate and flush the system, trapping foreign particles in the filter, without contaminating the servo valve, which was shipped clean from the factory.

The filter element should be inspected periodically (every one to two hours) during the flushing period to make sure that it is not clogged and the bypass is not opened. The filter element must be changed when it shows that it is becoming clogged.

The flushing period depends on overall system configuration and system contaminant level. When filter sampling shows that little or no additional contaminants are present, install a new filter element, remove the flushing plate, and install the servo valve.

Planned Maintenance

Establish periodic maintenance of the system. To provide good maintenance of a servo system, observe the following procedures:

1. Inspect and change filter elements at least every 500 hours or three months, whichever comes first.
2. Clean intake strainer to pump periodically.
3. Check the hydraulic fluid for acid or other contaminants. Odor can be a good indicator of fluid deterioration.
4. Repair leaks in the system.
5. Reflush the system as covered in a previous topic after you have added new oil or replaced parts that were not precleaned.
6. Ensure that no foreign particles are introduced into the system through the reservoir cap, oil filter plug, return line sealing gasket, or other openings in the reservoir.

If you adhere to the five points of preliminary flushing, as well as the six points of planned maintenance, you may expect virtually trouble-free operation of your servo system.

9.9   SUMMARY

1. The hydraulic system is but a link in the chain of power transfer from the prime mover to the load.
2. The energy may change form as it moves through a powered system but the power remains the same.

3. If there were no slippages in the power chain, there would be no need for velocity servo.
4. The best way to regulate speed is to use it to regulate itself (velocity servo).
5. A positional servo senses position to regulate itself.
6. Electronics makes the better "brain" and hydraulics the better "muscle."
7. Servo feedback signals can assume any one of many forms of energy: mechanical, hydraulic, pneumatic, or electrical.
8. Servo feedback signals may be fed back to any of the components in the chain.
9. The *best* method of feedback, however, is to feed back an *electrical* signal to the *hydraulic* part of the system.
10. The gain of the servo loop is limited by the lowest resonant frequency of the loop.
11. Low oil compressibility is a key to the response of the servo system.
12. No variable-speed power transmission exists today that can equal the stiffness—and therefore the accuracy—of an electrohydraulic servo system.
13. Pressure-compensated pumps are ideally suited for servo power supplies.
14. Piston motors are generally better servo actuators than the gear or vane types.
15. Leakage flow and breakaway pressure (required to generate the necessary breakaway force) are two important considerations in selecting servo cylinders.
16. A transducer performs the function of converting a source of energy from one form to another, such as hydraulics to electrical, or mechanical to electrical.
17. Depending on their function in the servo system, transducers may be classified as velocity, positional, pressure, acceleration, or flow types.
18. The heart of the electrohydraulic servo system is the servo valve.
19. Valve flow gain is the change in output flow per unit change in input current (slope of the flow curve).
20. Valve dynamics can be expressed either in terms of transient (step) response or sinusoidal frequency response.
21. A positional servo system, with an amplifier, can repeat a programmed position within ± 0.0003 in.
22. Most servo amplifiers are operational or use operational front ends. An operational amplifier, regulated with feedback, can have a very high gain. Without feedback, the operational amplifier would be unstable because of its excessive open-loop gain.
23. The voltage gain of the op amp is merely the ratio of R_o/R_i. So, to change the gain we need only to change the resistance of R_o.
24. The current gain of a servo amplifier is expressed as mA/V and is ideally suited to drive torque or force motors.
25. The integrating operational amplifier is the same as the linear operational type except that R_o is replaced with a capacitor.
26. When an integrating amplifier is used in a velocity system, velocity errors are changed to acceleration errors. So, give the amplifier enough time to integrate and it will reduce all velocity errors to zero.
27. The Kelvin–Varley voltage divider simulates a pot with 1000 steps.
28. The transfer function (or gain) of a device is the ratio of the output to the input.
29. A servo system automatically corrects its output to correspond to its input or command.

30. The primary advantage of a servo control system is that it provides accuracy while maintaining fast cycling time.
31. Open-loop gain is defined as the ratio of feedback signal to input (command) signal with the feedback signal disconnected from the summing point.
32. If the combined amplitude gain (total open-loop gain) is greater than 1 when the total phase shift is 360°, the system will oscillate.
33. A shortcut to estimating system performance requires that the sinusoidal response data be taken *only* of the component with the *lowest* resonant frequency.
34. Once the *resonant frequency is determined* for the cylinder, a *velocity constant can be assumed* for the system.
35. The total error of the system is determined by dividing the total deadband (mA) by the electrical gain (mA/in.).
36. The "cookbook" approach may be used to predetermine the approximate accuracy of many pot–pot positional systems.
37. The servo valve and cylinder may be sized with the use of Application Note 2.
38. Two systems used in metals processing are described in Section 9.6.
39. Servo trim procedures are covered in Section 9.7.
40. Servo maintenance is covered in Section 9.8.

9.10   PROBLEMS AND QUESTIONS

1. True or false?: Energy is consumed while going through a system.
2. True or false?: The parameters shown across the top of Figure 9.1 have their influence from left to right.
3. True or false?: The parameters shown across the bottom of Figure 9.1 have their influence from left to right.
4. True or false?: To change load speed, you may change any one or combination of parameters across the top of Figure 9.1.
5. True or false?: To regulate load speed completely, we must maintain constant speed under all changing load conditions.
6. True or false?: Each of the components of a system, with the exception of the gear box, will slip when the torque or pressure is increased.
7. True or false?: The best way to regulate speed is to use load compensation.
8. True or false?: The best way to regulate load speed is to sense speed and use it to regulate itself.
9. True or false?: The best method of servo control is to feed back a *hydraulic* signal to the *electrical* part of the system.
10. True or false?: To regulate the speed of the load, correction need only be provided for the slippage of the worst component of the system.
11. True or false?: If the load of a velocity servo slows down, the tachometer will generate less voltage.
12. True or false?: The gain of the servo loop is limited by the highest resonant frequency of the loop.
13. True or false?: Low oil compressibility is a key to the response of the electrohydraulic servo system.

14. True or false?: The higher the compressibility of the fluid, the higher the resonant frequency of the servo system.
15. True or false?: Pressure-compensated pumps are not well suited for servo power supplies.
16. True or false?: The piston-type motor generally has more internal leakage than the vane motor.
17. True or false?: The piston motor performs poorly in a servo system because it has a high breakaway pressure.
18. True or false?: Both the valve body and spool of the tracer valve are movable.
19. True or false?: The stylus of a tracer valve may be pushed into the valve by the force of your little finger.
20. True or false?: Electrohydraulic servo valves are always four-way and never three-way.
21. Assuming 100% efficiency throughout the system of Figure 9.1, what is the flow through the hydraulic motor if the hydraulic pressure is 1714 lb/in.^{2} and the electric input is 746 V at 10 A? (*Note:* One electric horsepower = 746 W, and volts × amperes = watts.)
22. What is the resonant frequency of a hydraulic valve–motor system if the displacement of the motor is 2 in.^{3}/rad, the volume of fluid under compression is 3.2884 in.^{3}, the compressibility of oil is 4.3×10^{-6} in.^{2}/lb, and the moment of inertia of the motor and reflected load is 10 lb-in.-sec^2?
23. What will be the acceleration ($a =$ in./sec^2) of a load if the force is ($f = 1000$ lb) and the mass is ($m = 100$ lb sec^2/in.)? Formula: ($a = \dfrac{f}{m}$).
24. What is the resonant frequency of a valve–cylinder system if the annular area of the cylinder is 3 in.^{2}, the length of the cylinder is 10 in., where 4 in.^{3} of oil is under compression, and the load weight is 1544 lb?
25. What is the flow through a meter motor rotating at 1000 rev/min if it has a displacement of 2.31 in.^{3}/rev? (*Note:* A meter motor is a motor calibrated in gal/min.)
26. True or false?: The frequency response of a single-stage servo valve is generally higher than a multi-stage design.
27. True or false?: When the spool of the four-way servo valve is at the center, a small flow is metered from the pressure port, past the A and B cylinder ports to tank.
28. True or false?: The hydraulic output flow from a servo valve is indirectly proportional to the amplitude of its electrical dc input current.
29. True or false?: Dither keeps the spool of a servo valve active through center.
30. True or false?: When a second stage is added to a servo valve, feedback must be added between the two stages.
31. True or false?: The no-amplifier servo system has excellent positional accuracy.
32. True or false?: The servo system with an amplifier has very good positional accuracy.
33. True or false?: The voltage gain of an operational amplifier is equal to R_o/R_i.
34. True or false?: The transfer function (or gain) of a device is the ratio of its output to its input.

35. True or false?: A servo system automatically corrects its output to correspond to its input or command.

36. True or false?: The amount of open-loop gain that can be used in a system is unlimited.

37. True or false?: The "cookbook" approach to servo system analysis is based on much testing and experience.

38. True or false?: Overshooting can be reduced or eliminated in a positional system with the use of a dual-gain circuit.

39. True or false?: The tracking error of a servo cylinder is inversely proportional to the cylinder speed.

40. True or false?: The best steady-state accuracy of a velocity servo system is obtained with an integrating amplifier.

41. True or false?: The best response of a velocity servo system is usually obtained with a linear amplifier.

42. True or false?: In the continuous casting system, the molten metal is siphoned from the ladle into a mold.

43. True or false?: The first step in trimming a system is to check for hydraulic leaks and proper phasing.

44. True or false?: If the cylinder goes hard over to either end when the system is first turned on, then reverse the excitation voltage to the command pot.

45. True or false?: Preliminary flushing is strongly recommended to ensure the removal of any contaminants from the system before installing the servo valve.

46. What would be the power rating of the pots of Figure 9.10 if the excitation voltage were raised to 15 V? (*Note:* The current through the torque motor, and therefore the pot, would be $E/R = 15/20 = 0.75$ A.)

47. What would be the power rating of the pots of Figure 9.11 if the excitation voltage were raised to 150 V?

48. What would be positional accuracy of the system of Figure 9.11 if the amplifier gain is set at 500 mA/V? (Assume the same valve deadband of 3 mA.)

49. What is the valve–cylinder resonance of the system of Figure 9.20 with the following assumptions?:

$$\text{Cylinder annulus area} = 7.5 \text{ in.}^2$$
$$\text{Volume of fluid under compression} = 90 \text{ in.}^3$$
$$\text{Load weight} = 4000 \text{ lb}$$
$$\text{Compressibility of oil } (E) = 5 \times 10^{-6} \text{ in.}^2/\text{lb}$$

50. What is the K_v allowed for the system of Problem 49?

51A. What is the recommended amplifier gain setting of the system of Problem 49 with the following additional assumptions?:

$$G_{sv} = 0.1 \ \frac{\text{in.}^3/\text{sec}}{\text{mA}} \quad H_{fb} = 5 \ \frac{\text{V}}{\text{in.}} \quad G_{cy1} = \frac{1}{7.5} \text{ in.}^2$$

51B. Determine the estimated accuracy of the system of Problems 49 thru 51A if the deadband is 5 mA.

52. Repeat Problems 49 through 51B with the following changes:

$$\text{Annulus area} = 10 \text{ in.}^2$$
$$\text{Volume of fluid under compression} = 100 \text{ in.}^3$$
$$\text{Load weight} = 3000 \text{ lb}$$
$$G_{sv} = 0.15 \text{ in.}^3/\text{mA}$$
$$G_{cy1} = 1/10 \text{ in.}^2$$
$$H_{fb} = 10 \text{ V/in.}$$

53. Using the instructions of Application Note 1 of the "cookbook" approach to accuracy predictions (Section 9.5), estimate the accuracy of a pot–pot system with the following parameters:

$$\text{Cylinder area} = 8 \text{ in.}^2$$
$$\text{Cylinder length} = 20 \text{ in.}$$
$$\text{Load weight} = 1000 \text{ lb}$$
$$\text{Load force} = 4000 \text{ lb}$$
$$\text{Supply pressure} = 1000 \text{ lb/in.}^2$$

54. Calculate the tracking error for a servo cylinder system with the following information known:

$$\text{Servo valve gain} = \frac{0.15 \text{ in.}^3/\text{sec}}{\text{mA}}$$
$$\text{Feedback gain} = 4 \text{ V/in.}$$
$$\text{Amplifier gain} = 800 \text{ mA/V}$$
$$\text{Cylinder gain} = 1 \text{ in.}/4 \text{ in.}^3$$
$$\text{Velocity of the slave cylinder} = 10 \text{ in./sec}$$

55. Calculate the closed-loop droop of a linear velocity servo system when the following information is known:

$$\text{Open-loop droop} = 25 \text{ rev/min (with load)}$$
$$\text{Feedback tach voltage} = 10 \text{ V with}$$
$$\text{Command voltage} = 15 \text{ V}$$

10

Troubleshooting Guide
and Maintenance Hints

Author's Note: Sections 10.1 and 10.2 are reproduced from Vickers Service Bulletin I-3998-S, revised September 1974, by permission of Vickers Inc., Maumee, Ohio.

I believe there is no better treatment of the subject matter than this document, which has been in use for more than 20 years by Vickers customers and personnel.

After the Vickers material is presented, comments, a summary, and problems and questions finish out the chapter.

See Section 9.8, Servo Maintenance, Chapter 9, for additional maintenance information.

10.1   INTRODUCTION

General

The trouble shooting charts and maintenance hints that follow are of a general system nature but should provide an intuitive feeling for a specific system. The more general information is covered in the immediately following paragraphs. Effect and probable cause charts appear in Section 10.2.

System Design

There is, of course, little point in discussing the design of a system which has been operating satisfactorily for a period of time. However, a seemingly uncomplicated procedure such as relocating a system or changing a component part can cause problems. Because of this, the following points should be considered:

1. Each component in the system must be compatible with and form an integral part of the system. For example, an inadequate size filter on the inlet of a pump can cause cavitation and subsequent damage to the pump.
2. All lines must be of proper size and free of restrictive bends. Undersize or restricted line results in a pressure drop in the line itself.
3. Some components must be mounted in a specific position with respect to other components or the lines. The housing of an in-line pump, for example, must remain filled with fluid to provide lubrication.
4. The inclusion of adequate test points for pressure readings, although not essential for operation, will expedite troubleshooting.

Knowing the System

Probably the greatest aid to troubleshooting is the confidence of knowing the system. Every component has a purpose in the system. The construction and operating characteristics of each one should be understood. For example, knowing that a solenoid-controlled directional valve can be manually actuated will save considerable time in isolating a defective solenoid. Some additional practices which will increase your ability and also the useful life of the system follow:

1. Know the capabilities of the system. Each component in the system has a maximum rated speed, torque, or pressure. Loading the system beyond the specifications simply increases the possibility of failure.
2. Know the correct operating pressures. Always set and check pressures with a gauge. How else will you know if the operating pressure is above the maximum rating of the components? The question may arise as to what the correct operating pressure is. If it isn't correctly specified on the hydraulic schematic, the following rule should be applied:

   The correct operating pressure is the lowest pressure which will allow adequate performance of the system function and still remain below the maximum rating of the components and machine.

   Once the correct pressures have been established, note them on the hydraulic schematic for future reference.
3. Know the proper signal levels, feedback levels, and dither and gain settings in servo control systems. If they aren't specified, check them when the system is functioning correctly and mark them on the schematic for future reference.

Developing Systematic Procedures

Analyze the system and develop a logical sequence for setting valves, mechanical stops, interlocks, and electrical controls. Tracing of flow paths can often be accomplished by listening for flow in the lines or feeling them for warmth. Develop a cause and effect troubleshooting guide similar to the charts appearing in Section 10.2. The initial time spent on such a project could save hours of system downtime.

Recognizing Trouble Indications

The ability to recognize trouble indications in a specific system is usually acquired with experience. However, a few general trouble indications can be discussed.

1. Excessive heat means trouble. A misaligned coupling places an excessive load on bearings and can be readily identified by the heat generated. A warmer than normal tank return line on a relief valve indicates operation at relief valve setting. Hydraulic fluids which have a low viscosity will increase the internal leakage of components resulting in a heat rise. Cavitation and slippage in a pump will also generate heat.
2. Excessive noise means wear, misalignment, cavitation or air in the fluid. Contaminated fluid can cause a relief valve to stick and chatter. These noises may be the result of dirty filters or fluid, high fluid viscosity, excessive drive speed, low reservoir level, loose intake lines, or worn couplings.

Maintenance

Three simple maintenance procedures have the greatest effect on hydraulic system performance, efficiency, and life. Yet, the very simplicity of them may be the reason they are so often overlooked. What are they? Simply these:

1. Maintaining a clean sufficient quantity of hydraulic fluid of the proper type and viscosity
2. Changing filters and cleaning strainers
3. Keeping all connections tight, but not to the point of distortion, so that air is excluded from the system.

10.2 TROUBLESHOOTING GUIDES

The following charts are arranged in five main categories. The heading of each one is an effect which indicates a malfunction in the system. For example, if a pump is exceptionally noisy, refer to Chart I, titled Excessive Noise. The noisy pump appears in Column A under the main heading. In Column A there are four probable causes for a noisy pump. The causes are sequenced according to the likelihood of its happening or the ease of checking it. The first cause is cavitation and the remedy is "a". If the first cause does not exist, check for cause number 2, etc.

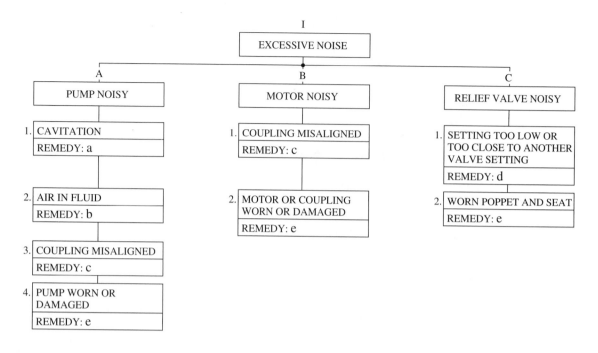

Excessive Noise Remedies

a. Any or all of the following: Replace dirty filters—Wash strainers in solvent compatible with system fluid—Clean clogged inlet line—Clean reservoir breather vent—Change system fluid—Change to proper pump drive motor speed—Overhaul or replace supercharge pump—Fluid may be too cold

b. Any or all of the following: Tighten leaky inlet connections—Fill reservoir to proper level (with rare exception all return lines should be below fluid level in reservoir)—Bleed air from system—Replace pump shaft seal (and shaft if worn at seal journal)

c. Align unit and check condition of seals, bearings and coupling

d. Install pressure gauge and adjust to correct pressure

e. Overhaul or replace

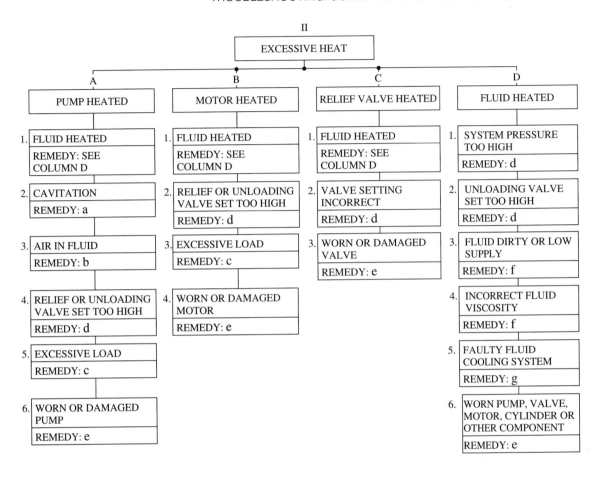

Excessive Heat Remedies

a. Any or all of the following: Replace dirty filters—Clean clogged inlet line—Clean reservoir breather vent—Change system fluid—Change to proper pump drive motor speed—Overhaul or replace supercharge pump

b. Any or all of the following: Tighten leaky inlet connections—Fill reservoir to proper level (with rare exceptions all return lines should be below fluid level in reservoir)—Bleed air from system—Replace pump shaft seal (and shaft if worn at seal journal)

c. Align unit and check condition of seals and bearings—Locate and correct mechanical binding—Check for work load in excess of circuit design

d. Install pressure gauge and adjust to correct pressure (keep at least 125 psi difference between valve settings)

e. Overhaul or replace

f. Change filters, and also system fluid if it is of improper viscosity—Fill reservoir to proper level

g. Clean cooler and/or cooler strainer—Replace cooler control valve—Repair or replace cooler

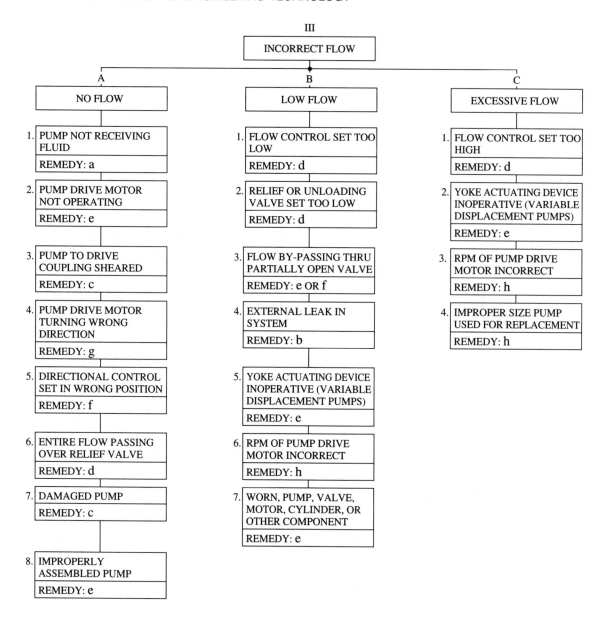

Incorrect Flow Remedies

a. Any or all of the following: Replace dirty filters—Clean clogged inlet line—Clean reservoir breather vent—Fill reservoir to proper level—Overhaul or replace supercharge pump

b. Tighten leaky connections—Bleed air from system

c. Check for damaged pump or pump drive—replace and align coupling

d. Adjust

e. Overhaul or replace

f. Check position of manually operated controls—Check electrical circuit on solenoid-operated controls—Repair or replace pilot pressure pump

g. Reverse rotation

h. Replace with correct unit

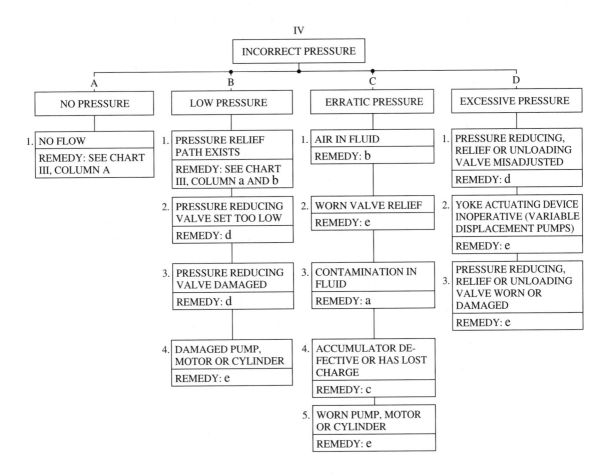

Incorrect Pressure Remedies

a. Replace dirty filters and system fluid

b. Tighten leaky connections (fill reservoir to proper level and bleed air from system)

c. Check gas valve for leakage—Charge to correct pressure—Overhaul if defective

d. Adjust

e. Overhaul or replace

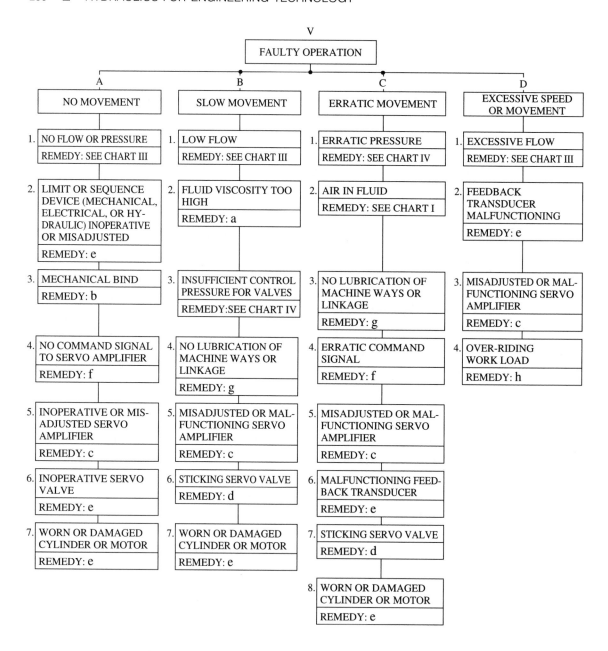

Faulty Operation Remedies

   a. Fluid may be too cold or should be changed to clean fluid of correct viscosity

   b. Locate bind and repair

   c. Adjust, repair, or replace

   d. Clean and adjust or replace—Check condition of system fluid and filters

   e. Overhaul or replace

  f. Repair command console or interconnecting wires

  g. Lubricate

  h. Adjust, repair, or replace counterbalance valve

10.3 COMMENTS

Bearing Life

A clean, well-maintained hydraulics system will operate within its specifications for many years. The most limiting factor of the well-maintained system is the bearing life expectancy of its rotating parts. Obtain this bearing-life information from the manufacturer and keep it in mind when considering the overhaul or replacement schedule of your pumps and motors.

Clean Oil

It would be wise to follow the flushing procedure described in Section 9.8 for all systems. Although this procedure is time consuming, it could more than pay for itself by avoiding downtime later.

Excessive Noise

Excessive noise usually means excessive wear. So try to operate the system at a pressure and speed range where the noise is at a minimum. If the system still sounds noisy, recheck Troubleshooting Guide I.

Excessive Heat

In many systems, the heat from the hydraulics is transferred to the product being processed. This could be a real problem when hamburgers or other meat products are being formed by hydraulically operated machinery. It may be necessary to add a water or air cooler for the oil rather than allow the product to cool the hydraulic equipment.

System Pressure and Flow

1. Do not expect system pressure (beyond what is necessary to push system flow through the pipes and dumped valves—10 to 20 $lb/in.^2$) without added restriction (such as load, relief, or other pressure-regulating valves).

2. Do not expect system flow without (at least) enough pressure to push the flow through the pipes and dumped valves.

10.4 SUMMARY

1. An inadequate filter (or strainer) on the inlet of the pump can cause cavitation and subsequent damage to the pump.

2. The housing of an in-line pump must remain filled with fluid in order to provide lubrication.
3. Adequate test points should be provided to facilitate pressure readings during troubleshooting.
4. The most important aid to troubleshooting may be your confidence of knowing the system.
5. The rule of thumb for setting the proper operating pressure is:

   The correct operating pressure is the lowest pressure which will allow adequate performance of the system function and will remain below the maximum rating of the components and machine.

6. Once the settings of system parameters have been determined, note them on the hydraulic drawing for future reference.
7. Analyze the system and develop systematic procedures for setting controls.
8. Learn to recognize trouble indications, such as noise and heat.

10.5   PROBLEMS AND QUESTIONS

1. True or false?: The most limiting factor of the well-maintained system is the bearing life expectancy of its rotating parts.
2. True or false?: Do not worry about dirty oil. The filters will take care of the dirt.
3. True or false?: Most solenoid-operated valves may be manually operated.
4. True or false?: It is good practice to "jog" the system for first startup after system repair.
5. True or false?: Excessive noise means excessive wear.
6. True or false?: Pressure is required to produce flow.
7. True or false?: Restriction is required to produce pressure.
8. True or false?: Excessive hydraulic motor noise may be caused by a misaligned motor coupling.
9. True or false?: Excessive pump noise may be caused by air in the fluid.
10. True or false?: Excessive relief valve noise may be caused by a worn poppet and seat.
11. Why do we sometimes use a larger sized cylinder in a servo system than is actually needed to move the load?

    (A) We want to increase the load speed.

    (B) We want to lower the acceleration of the load.

    (C) We want to increase the positional accuracy.

    (D) We want to move the load more easily through the "hard spots."

12. What would be the maximum extending speed of a 2:1 area cylinder in a *conventional fixed-flow circuit* if its maximum retracting speed is 100 in./min?

    (A) 50 in./min

    (B) 70.7 in./min

    (C) 100 in./min

   (D) 141.4 in./min

   (E) 144 in./min

   (F) 200 in./min

13. What would be the maximum extending speed of a 2:1 area *servo* cylinder if its maximum retracting speed is 100 in./min?

   (A) 50 in./min

   (B) 70.7 in./min

   (C) 100 in./min

   (D) 141 in./min

   (E) 200 in./min

14. What would be the recommended hp rating of the electric drive motor for a hydraulic system if the maximum flow is 15 gal/min at 1000 lb/in.^{2} the overall efficiency of the pump is 80%, and the electric motor is 80%?

   (A) 7.5 hp

   (B) 10 hp

   (C) 15 hp

   (D) 20 hp

   Calculate the following efficiencies for a hydraulic system with the inputs and outputs listed:

15. Electric drive motor overall efficiency.

16. Hydraulic pump overall efficiency.

17. System overall efficiency.

   Electric motor input power = 1.5 kW

   Electric motor output power = 1.5 hp

   Pump input power = 1.5 hp

   Pump output flow = 2 gal/min

   Pump output pressure = 1000 lb/in.^{2}

APPENDIX A
Glossary of Terms

Absolute pressure   The pressure above absolute zero. It is the sum of the atmospheric pressure and gauge pressure.

Absolute zero pressure   The reference zero that is not the atmospheric pressure. It is the pressure of a pure vacuum.

Accumulator   A container in which fluid is stored under pressure as a source of fluid power. May be spring- or air-pressure loaded.

Accuracy (Servo)   The ability of a system to achieve a desired output. Accuracy is sometimes expressed in terms of error. The smaller the error, the better the accuracy.

Actuator (hydraulic)   A device for converting hydraulic energy into mechanical energy (cylinder or motor).

Aeration   Air trapped in hydraulic fluid. Excessive aeration may cause the fluid to appear milky and the system to operate erratically.

Amplifier (electrical)   A device used to amplify (make larger) an electrical signal. The signal may be current, power, or voltage, and can be either ac or dc.

Analog device   A device that produces an output parameter whose amplitude is a function of an input parameter.

Annular area   A ring-shaped area. This usually refers to the effective area of the rod-side of a single-rod cylinder, i.e., the piston area minus the rod area.

Atmosphere (one)   The measure of pressure equal to about 14.5 lb/in.^{2} or 1 bar.

Atmospheric pressure   Pressure exerted by the atmosphere at any specific location on the earth's surface. Atmospheric pressure at sea level under normal barometric conditions is 14.7 lb/in.^{2} (absolute).

Attenuation (servo)   The ratio of the input of a device to its output. Opposite of gain (*see* Gain).

Baffle   A device, usually a plate, installed in a reservoir to separate flow of the return lines from the flow to the pump inlet.

Bar   An international standard pressure unit equal to approximately 14.5 lb/in.^{2} (average barometric pressure) or 100,000 Pa.

Bleed-off   To divert a specific controllable portion of pump delivery directly to tank.

Brake valve   A valve used in the exhaust line of a hydraulic motor to prevent an overrunning load and excessive pressure while decelerating or stopping the load.

Breakaway pressure or force   Pressure or force needed to start a load (usually a cylinder load) moving. This pressure or force is generally larger than the running load due to the "stiction friction."

Breather   A device that permits air to move into and out of a component to maintain atmospheric pressure.

Brush noise   Electrical noise caused when the carbon brushes of a motor or generator cross from one segment of a commutator to another (quite prevalent in tachometer generators).

Bypass   A secondary path for fluid or electric flow.

Cartridge (of a fluid filter)   The replaceable element.

Cartridge (of a vane pump)   Composed of rotor, ring, vanes, and one or both side plates.

Cartridge valve   A valve that is inserted into a standard-sized cavity of a manifold to control direction, pressure, or flow.

Cavitation   A localized gaseous condition within a liquid stream that occurs where the pressure is reduced to the vapor pressure level. Cavitation causes noise and excessive wear in a pump.

Charge (or supercharge)   (1) To replenish a hydraulic system above atmospheric pressure. (2) To fill an accumulator with fluid under pressure (*see* Precharge pressure).

Charge pressure   The gauge pressure at which replenishing fluid is forced into the system.

Check valve   A control valve that permits flow of fluid in one direction only.

Choke   A restriction, the length of which is long compared to its cross-sectional dimension.

Circuit   An arrangement of hydraulics, pneumatics, or electric components interconnected to form a specific function within a system.

Closed-center switch (electrical)   A switch in which one or more paths are free to pass electric current in the center or neutral condition.

Closed-center valve (hydraulic or pneumatic)   A valve in which all ports are blocked in the center or neutral condition.

Closed circuit (electrical)   A circuit in which a complete path is provided for the flow of electric current.

Closed circuit (hydraulic)   A circuit arrangement in which pump output, after passing through other components, returns directly to pump inlet (instead of to tank).

Closed contacts (of an electric switch)   Contacts between which electric current may flow freely.

Closed loop (servo)   A system in which the feedback of one or more elements (hydraulic, electrical, pneumatic, or mechanical) is compared to a command signal, providing an error signal to control the output of the loop.

Closed ports (of a hydraulic valve)   Ports of the valve which are blocked and do not permit flow.

Command signal (input signal to a servo loop)   An external signal representing a new pressure, flow, position, velocity, acceleration, or other parameter to which the servo must respond.

Compensator control   An automatic displacement control for variable pumps and motors. It alters the displacement in response to system pressure changes as related to the adjusted pressure setting.

Component   A single hydraulic, electrical, pneumatic, or mechanical unit.

Compressibility   The change of volume of a unit volume of fluid when it is subjected to a unit change of pressure. *Note:* The compressibility of liquids (such as oil) is very small compared to the compressibility of gases (such as air).

Controllability   The finest adjustable increment of a system.

Cooler   A heat exchanger that removes heat from a system.

Counterbalance valve   A pressure control valve that maintains back-pressure to prevent a load from falling. The valve is usually connected to the output of a vertical double-acting cylinder to support weight or prevent uncontrolled falling or dropping.

Cracking pressure   The pressure at which a pressure-operated valve begins to pass fluid.

Current (electrical)   The flow from one point to another around a closed electrical circuit. Current is measured in amperes, coulombs per second, or volts/ohm.

Cushion   A device that provides controlled resistance to motion. The cushion can be built into either end of a hydraulic cylinder, thus restricting the output flow and providing a cushion to the load near the end of travel.

Cycling   A rhythmic change of the parameter under control.

Cylinder   A device that converts fluid power (pressure times flow) into linear mechanical power (force times speed). A cylinder usually consists of a movable piston and rod operating within a cylindrical bore to move a linear load.

Dancer pot   A potentiometer mounted to a spring, or gravity-loaded arm connected to a reel, which rides the web of material being processed. As the pot rotates, its output voltage measures the web tension against the arm.

Deadband (servo)   The region or band of no response where an error signal will not cause a corresponding actuation of the controlled variable.

Decompression   The slow release of confined fluid to gradually reduce pressure on the fluid.

Delivery   The volume of fluid discharged by a pump in a given time. Delivery is usually expressed as gallons per minute (gal/min) or liters per minute (L/min).

Demodulator   A device that is used to retrieve the signal from a modulated carrier. Amplitude demodulators convert ac or amplitude modulated waves to dc or low-frequency signal voltages.

Devent   To close the vent connection of a pressure control device, allowing the device to function at its adjusted pressure setting.

Differential cylinder   Any cylinder in which the two opposed piston areas are not equal.

Digital signal (electrical)   A voltage or current that varies (almost instantaneously) between two distinct and fixed levels.

Diode (electrical)   An electrical device that allows current to flow in one direction only. The hydraulic analogy of a diode is a check valve.

Diode rectifier (electrical)   A device used to convert ac to dc.

Direct current (dc)   A steady level of electrical current that flows only in one direction in a circuit. DC current can be produced by a generator, battery, thermocouple, photocell, or by rectifying ac.

Directional valve   A valve whose primary function is to direct or prevent flow through selected channels.

Displacement (electrical pots, etc.)   Angular rotation of the wiper of a rotary pot or linear movement of the wiper of a linear pot.

Displacement (hydraulic)   The volume of fluid delivered by a pump or required by a motor during one revolution or (in some cases) during one radian ($1/2\pi$ revolutions). For example, 6.28 in.^{3}/rev or 1 in.^{3}/rad.

Dither (servo)   A low-amplitude, high-frequency signal inserted into a servo loop to minimize the effect of coulomb friction, hysteresis, and deadband. The secret to a well-adjusted dither is that all components are responding to the dither signal except the load.

Double-acting cylinder   A cylinder in which hydraulic force may be applied to the movable element in either direction.

Drain   A passage in, or a line from, a hydraulic component that returns leakage flow directly to tank or vented manifold.

Drift (servo amplifier)   The percentage above or below the programmed operating level of a servo amplifier over a specified operating time.

Dynamic behavior (servo)   Describes how a control system or component reacts, with time, when subjected to a changing input signal.

Dynamic error (servo)   The error that exists during a transient condition, when the system is moving from one steady state to another.

Efficiency   The ratio of the output of a device to its input (using the same units for both the input and output). For example, the overall efficiency of a pump is the output horsepower divided by the input horsepower. Exception to the output-over-input rule: The volumetric efficiency of a pump is the *actual* output in gal/min divided by the *theoretical* or design output in gal/min. Efficiency is usually expressed as a percent.

Electrohydraulic servo valve (servo)   A type of servo valve, most of which are capable of providing an output flow directly proportional to the input electric current. In a four-way servo valve the output flow is nearly zero when the input current is zero. It provides flow in one direction with one electrical polarity, and reverses flow when the electrical polarity is reversed.

Electromotive force (EMF) (electrical)   The force provided by the difference of electrical potential that causes current to flow in an electric circuit. The hydraulic analogy to EMF is pressure drop.

Energy   The ability or capacity to do work, measured in units of work (ft-lb or volt-coulomb).

Error (servo)   The amount by which a control system misses its target. For example, (1) How close is the load of a positional servo to its programmed position? (2) How close is the speed of the velocity servo system to the programmed speed?

Error signal (servo)   The signal that is the algebraic summation of an input signal and the feedback signal.

Feedback loop (servo)   Any closed loop consisting of one or more feedback elements that transmit a signal to a summation point.

Feedback signal (servo)   The signal from a feedback transducer.

Feedback transducer (servo)   A transducer is a device that converts one source of energy to another. The feedback transducer of an electrohydraulic servo system converts the controlled output to an electrical signal. Examples of feedback transducers: tachometer generator (velocity), feedback pot (positional), linear variable differential transducer (LVDT) (positional).

Filter (hydraulic)   A device used to separate and retain insoluble contaminants from a fluid.

Flow control valve   A valve whose primary function is to control or regulate flow rates.

Flow rate   The volume (gal, in.^{3}, or liter) or weight (lb) of fluid passing a given point in a conductor, per unit of time (minute).

Flying saw   A powered saw which cuts a moving material by first catching up and traveling at the speed of the material while making the cut.

Follow valve (servo)   A servo valve with a mechanical input and a mechanical feedback from the load cylinder such that the load position follows the mechanical input to the valve.

Force   Any push or pull measured in units of weight. In hydraulics, force may be determined by:

$$\frac{\text{lb (pressure)}}{\text{in.}^2} \times \text{in.}^2 \text{ (area)} = \textbf{lb}$$

$$\frac{\text{Newtons}}{\text{m}^2} \times \text{m}^2 = \text{newtons}$$

Four-way valve   A directional valve having four flow paths.

Frequency   The number of times an action occurs in a unit of time (cycles/second). The unit of cycles/second = Hertz.

Natural frequency   (*See* resonant frequency) Any object will vibrate at its natural (resonant) frequency when shocked or struck a hard blow (for example, a tuning fork). See Chapter 9 for determining the natural frequency of an electrohydraulic servo system.

Noise frequency   Frequency is the basis of all sound (audio frequency). The basic noise frequency of a pump or motor is equal to its speed in revolutions per second multiplied by the number of pumping chambers. For example, the basic noise frequency of a nine-piston pump driven at a speed of 1800 rev/min would be:

$$\frac{\cancel{min}}{\cancel{60}\ sec} \times \frac{\overset{30}{\cancel{1800}}\ \cancel{rev}}{\cancel{min}} \times \frac{9\ cycles}{\cancel{rev}} = 270\ cycles/sec\ (or\ 270\ cps\ or\ 270\ Hz)$$

Frequency response analysis (servo)   The response of a control system can be tested by applying a variable-frequency input signal (such as an ac voltage) and measuring the output. With this method, it is possible to predict system performance in advance.

Full-flow filter   A hydraulic filter where all the flow must pass through the filter elements.

Full-flow pressure   Pressure at full flow.

Gain (servo)   The ratio of the output signal of a device to its input. The opposite of attenuation (*see* Attenuation).

Gauge pressure   A pressure measured by a gauge, or calculated, using the atmosphere as zero reference.

Ground (electrical)   A point of zero reference in an electrical circuit, to which all other voltages are compared. This reference point is usually connected to earth ground for safety reasons.

Head pressure   Pressure caused by the weight of the fluid in a hydraulic system, or fluid density [weight per unit volume (lb/in.^{3})] times the height of the fluid (in.)

$$Example:\quad \frac{lb}{\cancel{in.}^{\cancel{3}}_{\ in.^2}} \times \cancel{in.} = \textbf{lb/in.}^{\textbf{2}}$$

Heat   A form of energy that causes warmth in a body. One Btu (British thermal unit) is the energy required to raise the temperature of one pound of water by one degree Fahrenheit.

Heat exchanger (liquid-to-air)   A device that removes heat from a hydraulic system by the use of an air fan.

Heat exchanger (liquid-to-liquid)   A device that removes heat from a hydraulic system by the use of water or some other liquid in a circulating system.

Hertz   The unit of frequency in cycles per second. 1 cps = 1 hertz.

Horsepower (hp)   The power required to lift 550 lb one foot in one second or 33,000 pounds one foot in one minute. A horsepower is also equal to 746 W (electrical) or 42.4 Btu per minute (heat).

Hunting (servo)   The tendency of a servo system to oscillate (move back and forth) continuously.

Hydraulic balance   A condition of equal and opposed hydraulic forces acting on a part in a hydraulic system.

Hydraulic control   A control that is actuated by hydraulically induced forces.

Hydraulics   The science and application of fluid power, pertaining specifically to fluid pressure, flow, and restriction.

Hydrodynamics   Engineering science pertaining to the energy of liquid flow and pressure.

Hydrokinetics   Engineering science pertaining to the energy of liquids in motion.

Hydropneumatics   Engineering science pertaining to the combination of hydraulic and pneumatic fluid power.

Hydrostatics   Engineering science pertaining to the energy of liquids at rest.

Hysteresis (servo)   The difference between the response to an increasing signal and that to a decreasing signal having the same slope.

Impedance (electrical)   The total opposition of a circuit to ac, including resistance, inductive reactance, and capacitive reactance.

Input signal (servo)   *See* Command signal.

Insulator (electrical)   A material that blocks the flow of electric current.

Joule   A unit of work or energy equal to one Newton-meter (N-m).

Kinetic energy   Energy that a substance or body has by virtue of its mass (weight) and velocity.

Lag (servo)   Engineering term for delay in response (usually in degrees).

Laminar flow   A flow condition where flow particles move in continuous parallel paths; streamlined flow (no turbulence).

Lands   Portion of valve spool that seals flow from undesired passages while permitting flow to desired ports.

Leverage   A gain in output force over input force by sacrificing the output distance moved. Mechanical advantage or force multiplication.

Lift   The height a body or column of fluid is raised, for instance, from the reservoir to the pump inlet. Lift is sometimes used to express a negative pressure or vacuum (the opposite of head).

Line (hydraulic)   A hose, pipe, or tube used to conduct hydraulic flow.

Linear actuator   A device for converting hydraulic energy into linear movement—a cylinder or ram.

Linearity (servo)   (of a servo valve, for instance) The degree of straightness of the hysteresis plot, or, how closely the output of a device follows the input.

Liter (metric)   A unit of volume. One liter = 1000 cm^3 and one liter = 0.264 gal (approximately).

Logic circuit   A digital circuit (or gate) with binary inputs and outputs and capable of performing decision-making functions.

Manifold   A fluid conductor with many connection ports.

Manual control   A control actuated by the human operator.

Manual override   A means of manually actuating an automatically controlled device.

Maximum pressure valve   *See* Relief valve.

Mechanical control   A control that may be actuated by gears, cams, linkages, screws, or other mechanical devices.

Meter   *Verb:* To regulate the amount or rate of flow. *Noun:* An instrument (with digital or dial readout) used to measure the quantity or value of a parameter.

Meter or metre (metric)   Unit of linear measurement. One metre = 39.37 in. = 1,000,000 μm = 1000 mm = 100 cm = 10 decimeters.

Micron   Unit of linear measurement. One micron (or micrometer) = one millionth of a meter or about 0.00004 in.

Micron rating   The smallest size particle, measured in microns, that a filter will remove.

Microprocessor (servo)   A digital computer consisting of a single miniature integrated circuit that has programmability and computational ability. Microprocessors are used in electrohydraulic as well as electromechanical systems.

Modem   Short for *mod*ulator/*dem*odulator. A device used for digital data communication between two separate computer processing units. The modulator converts digital information into tones that are transmitted over the telephone lines, and the demodulator changes the received tones back into digital data.

Motor (electrical)   A device that converts electrical energy to rotary mechanical energy.

Motor (hydraulic)   A device that converts hydraulic energy to rotary mechanical energy.

Multimeter (electrical)   A device (VOM or DMM) used to measure electrical quantities (volts, amps, ohms, watts, etc.).

Newton   Metric unit of force equal to 0.225 lb(*f*). One Newton is the hydraulic force required over an area of one square meter to produce a hydraulic pressure of one Pascal.

Noise (brush)   *See* Brush noise.

Noise (electrical)   Interference caused by unwanted signals superimposed onto normal control signals.

Null (servo)   The condition where a servo amplifier, servo valve, or other active servo device provides zero output.

Ohm's law (electrical)   The electrical expression of "Result is directly proportional to effort and inversely proportional to opposition." Current equals Voltage divided by Resistance, or $I = E/R$, where I = current intensity in amperes, E = electromotive force in volts, and R = resistance in ohms.

    The hydraulic equivalent of Ohm's law is "Restriction is directly proportional to Pressure and inversely proportional to Flow." Or $R =$

$$\frac{\text{lb/in.}^2}{\text{gal./min}}$$

where R = restriction in Lohms.

Open center circuit   A circuit condition where the pump output flow recirculates freely to reservoir, or back to the pump inlet, when the directional control valve is in its center or neutral position.

Open-centered valve   A valve in which all ports are interconnected in the center or neutral position.

Open circuit (electrical)   A circuit in which a complete electric path does not exist.

Open circuit (hydraulic)   A circuit in which a complete hydraulic path does exist.

Operational amplifier (op amp) (servo)   An electronic amplifier (usually integrated) with such high gain and input impedance that it is seldom used without feedback. The op amp is very versatile. With the proper feedback loop, it can produce any function needed. It is especially suitable for electrohydraulic servo systems.

Orifice   *See* Restricter, orifice.

Overshoot (servo)   The extra distance that a load travels or extra speed that exists during the transient state of changing position or velocity.

Passage   A machined or cored path through a hydraulic component to allow flow passage.

Phase shift (servo)   The time difference between the input and output signals of a control unit or system. The units of delay are usually degrees.

Pickling   An acid treatment of pipes, tubing, and fittings to remove scale that could contaminate hydraulic systems. Pickling procedures are described in Army and Navy maintenance manuals as well as in some hydraulic manufacturers' pamphlets.

Pilot pressure   Auxiliary pressure used to actuate or control other hydraulic components.

Pilot valve   The controlling valve of a two- or three-stage valve or any auxiliary valve used to operate another valve.

Piston   The movable part of a cylinder that fits inside the cylindrical bore. A rod, smaller than the piston, connects the piston to the movable load.

Plunger (ram)   The movable part of a cylinder that fits inside the cylindrical bore. The rod connecting the piston to the load is the same size as the piston.

Poppet   The part of a valve that blocks flow when it closes against a seat.

Port   The beginning or end of a passage in a component.

Positive displacement   The characteristic of a pump or motor in which the inlet is positively sealed from the outlet. Since the fluid cannot readily circulate within the unit, the rate of fluid delivery is virtually constant with rotation.

Pot   *See* Potentiometer.

Potentiometer (pot) (servo)   A device that produces a voltage versus displacement when excited by ac or dc. It is used as a feedback transducer or to provide an adjustable voltage.

Power   The rate of doing work. It is measured in ft × lb/min, horsepower, or watts.

Precharge pressure   The pressure of compressed gas in an accumulator prior to the admission of liquid.

Pressure   Force per unit area expressed in $lb/in.^2$ or N/m^2 (Pascals). Pressure is also expressed as heads of liquids, such as inches of mercury (Hg) or feet of water. Pressure may also be expressed in atmospheres or bars (1 bar = 14.5 $lb/in.^2$ or 100 kPa).

Pressure, breakaway *See* Breakaway pressure.

Pressure differential (or differential pressure) The difference in pressure between any two points in a system.

Pressure drop The pressure differential across a component or system.

Pressure override The difference between the cracking pressure of a valve and the pressure reached when the valve is passing full flow.

Pressure-reducing valve A valve that lowers and regulates the lower output pressure.

Pressure switch An electric switch turned on and off by fluid pressure.

Proportional-flow filter A filter designed to pass part of the flow through the filter element proportional to the pressure drop.

Proportional valve A valve that allows remote proportional control of pressure, flow, acceleration, and direction. It has most of the characteristics of a servo valve, but is not as responsive.

Pump, hydraulic A device that converts mechanical force and motion into hydraulic fluid power.

Ram A single-acting cylinder with a single-diameter plunger rather than a piston and rod (*see* Plunger).

Reciprocation Back-and-forth straight-line motion or oscillation.

Regenerative circuit (electronic or servo) A closed-loop positive feedback system in which the feedback adds to the error, thus sustaining continuous oscillations.

Regenerative circuit (hydraulic) A circuit arrangement for a differential cylinder in which discharge fluid from the rod end is combined with the pump delivery to be directed into the cap end.

Relay (electrical) An electromagnetically operated switch that allows one electric circuit to turn on or off another circuit.

Relay logic An electrical circuit of various interconnected relays used for controlling a machine or process based on the status of the relays.

Relief valve A pressure control valve whose primary function is to limit system pressure.

Repeatability (servo) The exactness with which motion or position can be duplicated.

Replenish To add fluid in order to maintain a full hydraulic system.

Reservoir A container for storage of liquid in a fluid power system.

Resistance (electrical) The opposition to current flow offered by components of an electric circuit.

Resistor (electrical) A component designed to offer resistance to an electrical current flow.

Resonant frequency The natural frequency of a tuning fork; or the natural frequency of any system that has been shocked into vibration.

Restriction (hydraulic) The opposition to hydraulic flow offered by components of a hydraulic circuit.

Restrictor   A component designed to offer restriction to a hydraulic flow.

Restrictor, choke   A restrictor the length of which is relatively large with respect to its cross-sectional area.

Restrictor, orifice   A restrictor the length of which is relatively small with respect to its cross-sectional area.

Return line   A line used to carry exhaust fluid from working drives to reservoir.

Reversing valve   A four-way directional valve used to reverse a double-action cylinder or reversible motor.

Rotary actuator   The name reserved for a hydraulic motor that has less than 360° rotation.

Sensitivity (servo)   The minimum input signal to a device required to produce a detectable output signal.

Sequence   The order of a series of operations or movements of a device.

Sequence valve   A pressure-operated valve, which at a predetermined pressure setting, diverts flow to a secondary circuit while holding the predetermined minimum pressure on the first circuit.

Servo   Using feedback to obtain automatic control.

Servo mechanism   A servo loop where the output load tracks as if it were driven by the input signal, even when the output power is many times that of the input. The output power is derived from the main or auxiliary power source and not from the servo input.

Servo system   A servo mechanism or any other closed-loop system where the output position, speed, pressure, torque, or power is automatically controlled by the feedback signal as a function of the input signal.

Servo valve   A follow valve or a valve that modulates output flow and pressure as a function of input electric current.

Signal (servo)   A command or request for a desired position, speed, torque, acceleration, flow, etc.

Single-acting cylinder   A cylinder in which hydraulic force may act in only one direction. The cylinder must be returned by gravity, spring, or other mechanical force.

Sinusoidal signal or wave   A sine wave signal. A signal or wave whose displacement (or amplitude) is the sine or cosine of an angle proportional to time or distance or both.

Slip   Caused by internal leakage of fluid.

Solenoid (electrical)   *Original usage:* A single-coil device—not a transformer with many coils. *Present usage:* An electromechanical device that converts electrical energy into linear mechanical motion.

Spool   A term loosely applied to any moving cylindrically shaped part of a valve that directs flow through the system.

Stability (servo)   The ability of a system to maintain control during outside disturbances.

Static behavior (servo)   Describes how a system or component behaves under fixed conditions. (As opposed to dynamic behavior, which describes how a system or component behaves under changing conditions.)

Step change   The change from one value to another in a single step.

Stiction friction   The friction that causes extra pressure and force (in addition to the acceleration force and pressure) to start the load moving.

Strainer   A coarse filter.

Streamline flow   See Laminar flow.

Stroke   *Noun:* The length of travel of a piston or plunger. *Verb:* To change the displacement of a hydraulic pump or motor.

Subplate   An auxiliary ported plate for mounting hydraulic components.

Suction line   The hydraulic line connecting the inlet of a pump to a reservoir.

Sump   *See* Reservoir.

Supercharge   To replenish a hydraulic system above atmospheric pressure.

Surge pressure   Increased pressure due to dynamic conditions in the system.

Swash plate   A stationary canted plate in an axial-type piston pump that causes the pistons to reciprocate as the cylinder block and pistons rotate.

Synchro (servo)   A rotary electromechanical device used as an ac position feedback or reference signal.

Tachometer (servo)   A rotary device that generates ac or dc voltage proportional to speed and polarity or phase to denote the direction of rotation. The tachometer is used as feedback for velocity systems.

Tank   Reservoir or sump.

Throttle   Used to restrict flow. May be used as a noncompensated flow control or to produce a pressure drop as part of a compensated flow valve.

Torque   Rotary thrust or turning effort of a motor or other rotary device. Torque is usually expressed as lb-in. to avoid confusion with work, which is expressed as in.-lb.

Torque converter   A rotary fluid coupling that is capable of increasing torque.

Torque motor (servo)   A rotary electromechanical transducer used to actuate servo-valves.

Transducer (servo)   See Feedback transducer.

Transfer function   A mathematical expression of the output to input signals of a control. For examples, (1) the transfer function of a servo amplifier is [(output) mA/(input)V] = **mA/V** and (2) the transfer function of a cylinder is [(output) in./ (input) in.^{3}] = $1/\text{in.}^2$.

Transformer (electrical)   Transfers ac energy from one coil of conductive wire to another through electromagnetic coupling, without direct electrical contact.

Turbine   A device that is rotated by the impact of moving fluid against blades or vanes.

Turbulent flow (or turbulence)   A flow condition where flow particles move in a random manner rather than in parallel paths as in laminar flow. The usual cause of turbulence is the use of sharp-edged orifices in the control valves.

Unload   To unload pressure by dumping the flow directly to tank. Unloading saves energy during standby.

Unloading valve Allows the pressure of one circuit to be unloaded when the pressure of another circuit reaches a predetermined setting.

Vacuum Pressure below atmospheric pressure. Usually expressed in inches of mercury (in. Hg) as referenced to the existing atmospheric pressure.

Valve A valve that controls fluid flow, pressure, or direction.

Vapor pressure The pressure of a fluid (usually lower than atmospheric) at which the vapor separates from the solution. For example, when the hydraulic pump inlet pressure gets too low, air comes out of solution, causing pump cavitation, which results in excessive noise and pump wear.

Velocity (1) The speed of fluid in a line; in U.S. units, inches per second, inches per minute, or feet per minute; in SI units, meters per minute, centimeters per minute, or millimeters per second. (2) The speed of a linear motion; in U.S. units, inches per second, inches per minute, feet per minute, or miles per hour; in SI units, millimeters per second, centimeters per minute, meters per minute, or kilometers per hour. (3) The speed of rotary motion in radians per second or revolutions per minute.

Vent To connect to atmosphere.

Viscosity A measure of the internal resistance of a fluid or the resistance of a fluid to flow.

Volume The size of a chamber or space in gallons, liters, in.^{3}, or cm^2.

Work (electrical) An emf of one Volt moving one coulomb is one volt-coulomb of work. (See *Joule* for metric unit of work.)

Work (mechanical) Force acting through a distance. Lifting one pound one foot is performing one ft-lb of work.

APPENDIX B
Typical Symbols Used in Hydraulic Circuits—Alphabetized

A	ACCUMULATOR, GAS CHARGED	
	ACCUMULATOR, SPRING LOADED	
C	CHECK VALVE, DIRECT	
	CHECK VALVE, PILOT OPERATED	
	COOLER	
	CYLINDER, SINGLE ENDED NO CUSHIONS	
	CYLINDER, DOUBLE ENDED NO CUSHIONS	
	CYLINDER, SINGLE ENDED HEAVY-DUTY ROD, NO CUSHIONS	
	CYLINDER, SINGLE ENDED, ADJUSTABLE CUSHION, ROD END	
	CYLINDER, SINGLE ENDED, ADJUSTABLE CUSHION, CAP END	
	CYLINDER, SINGLE ENDED, ADJUSTABLE CUSHIONS, CAP AND ROD ENDS	

C (cont'd.)

COUNTERBALANCE VALVE WITH INTEGRAL CHECK	
DECELERATION VALVE, NONADJUSTABLE	D
DIRECTIONAL VALVE TWO-POSITION TWO-WAY	
DIRECTIONAL VALVE TWO-POSITION THREE-WAY	
DIRECTIONAL VALVE TWO-POSITION FOUR-WAY	
DIRECTIONAL VALVE TWO-POSITION IN TRANSITION	
DIRECTIONAL VALVE THREE-POSITION FOUR-WAY	
DIRECTIONAL VALVE WITH INFINITE POSITION ABILITY INDICATED BY HORIZONTAL BARS ON SIDES	
DIRECTIONAL VALVE, NO SPRING, DETENTED, MANUALLY OPERATED	

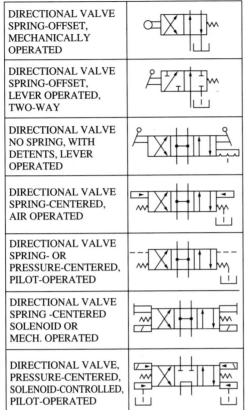

D (cont'd.)		
	DIRECTIONAL VALVE SPRING-OFFSET, MECHANICALLY OPERATED	
	DIRECTIONAL VALVE SPRING-OFFSET, LEVER OPERATED, TWO-WAY	
	DIRECTIONAL VALVE NO SPRING, WITH DETENTS, LEVER OPERATED	
	DIRECTIONAL VALVE SPRING-CENTERED, AIR OPERATED	
	DIRECTIONAL VALVE SPRING- OR PRESSURE-CENTERED, PILOT-OPERATED	
	DIRECTIONAL VALVE SPRING -CENTERED SOLENOID OR MECH. OPERATED	
	DIRECTIONAL VALVE, PRESSURE-CENTERED, SOLENOID-CONTROLLED, PILOT-OPERATED	

		D (cont'd.)
DIRECTIONAL VALVE SOLENOID-CONTROLLED, PILOT-OPERATED		
DIRECTIONAL VALVE SPRING-CENTERED, SOLENOID-CONTROLLED, PILOT-OPERATED WITH MANUAL OVERRIDE	OR	

FLOW VALVES F

FLOW CONTROL ADJUSTABLE NONCOMPENSATED	
FLOW CONTROL ADJUSTABLE NONCOMPENSATED WITH REVERSE FREE FLOW	
FLOW CONTROL WITH PRESSURE AND TEMPERATURE COMPENSATION	
FLOW CONTROL WITH PRESSURE AND TEMPERATURE COMPENSATION WITH REVERSE FREE FLOW	
FLOW METER	

F (cont'd.)	FLOW CONTROL REMOTELY CONTROLLED, ELECTRICALLY MODULATED WITH CHECK VALVE		LINE, PLUGGED PORT, OR TESTING MEASUREMENT, OR POWER TAKE-OFF		L (cont'd.)
H	HEATER		LINE, TO RESERVOIR ABOVE FLUID LEVEL BELOW FLUID LEVEL		
L	LINE, CROSSING	OR	LINE, WITH FIXED RESTRICTION		
	LINE, DRAIN, LIQUID		LINE, WORKING (MAIN)		
	LINE, ENCLOSURE COMPONENT		METHODS OF OPERATION		
	LINE, FLEXIBLE		DETENT		M
	LINE, FLOW HYDRAULIC		MANUAL		
	LINE, FLOW PNEUMATIC		MECHANICAL		
	LINE, JOINING		PILOT PRESSURE REMOTE SUPPLY INTERNAL SUPPLY		
	LINE, PILOT (FOR CONTROL)		PEDAL OR TREADLE		

M METHODS OF OPERATIONS (cont'd.)

PRESSURE COMPENSATED		PRESSURE RELIEF VALVE	P (cont'd.)
PUSH BUTTON		PRESSURE RELIEF VALVE, REMOTE ELECTRICALLY MODULATED	
PUSH–PULL LEVER		PRESSURE RELIEF VALVE, UNLOADING WITH INTEGRAL CHECK	
SERVO MOTOR			
SOLENOID, SINGLE WINDING		PRESSURE SWITCH	
SPRING		PROPORTIONAL VALVE	

MOTORS

MOTOR ELECTRIC		PUMP FIXED DISPLACEMENT, WITH DRAIN, UNIDIRECTIONAL	
MOTOR HYDRAULIC FIXED DISPLACEMENT, DUAL DIRECTIONAL			
MOTOR HYDRAULIC VARIABLE DISPLACEMENT, DUAL DIRECTIONAL		PUMP FIXED DISPLACEMENT, WITHOUT DRAIN, UNIDIRECTIONAL	
P PRESSURE INDICATOR			
PRESSURE-REDUCING VALVE			

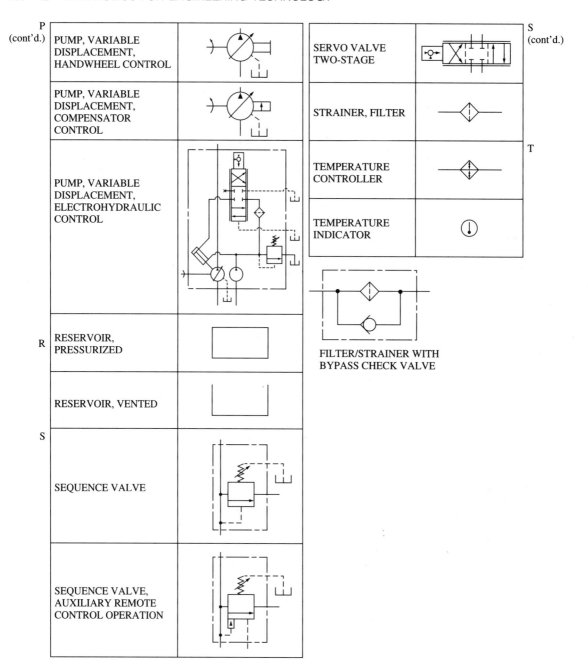

P (cont'd.)	PUMP, VARIABLE DISPLACEMENT, HANDWHEEL CONTROL	
	PUMP, VARIABLE DISPLACEMENT, COMPENSATOR CONTROL	
	PUMP, VARIABLE DISPLACEMENT, ELECTROHYDRAULIC CONTROL	
R	RESERVOIR, PRESSURIZED	
	RESERVOIR, VENTED	
S	SEQUENCE VALVE	
	SEQUENCE VALVE, AUXILIARY REMOTE CONTROL OPERATION	

S (cont'd.)	SERVO VALVE TWO-STAGE	
	STRAINER, FILTER	
T	TEMPERATURE CONTROLLER	
	TEMPERATURE INDICATOR	

FILTER/STRAINER WITH BYPASS CHECK VALVE

APPENDIX C
Solutions to Odd-Numbered Problems and Questions

Chapter One

1. List two units of pressure.

   Answer: lb/in.^{2} (psi) and Pascals.

3. True or false?: Head pressure increases with volume if the height of the fluid is held constant.

   Answer: False. Head pressure increases only with height.

5. List three cases where head pressures are significant.

   Answer: 1. Where the other pressures involved are very low, such as at the inlet of a pump,
   2. Where the height of the fluid is great, such as at the bottom of the ocean, or
   3. Where the area affected by the head pressure is very large.

7. What would have been the total pressure, including the head pressure, at a point 2 in. from the bottom of the jug of Figure 1.3b if the cork were pushed down with a force of 10 lb?

   Answer: The pressure at any point on the inside surface of the jug caused by the 10-lb weight, acting on the 1 in.^{2} area of the cork, would be 10 lb/in.^{2}. In addition, a head pressure will exist at any point caused by the height of the fluid above that point. Thus the total pressure at a point 2 in. from the bottom of the jug would be 10 lb/in.^{2} plus the head pressure of 12 in. minus 2 in. or 10 in. The pressure caused by the 10 in. of head would be

$$(10 \text{ in.}/12 \text{ in.}) \times (0.4 \text{ lb/in.}^2) = 0.3333 \text{ lb/in.}^2$$

So the total pressure at the 2-in. point is

$$10 + 0.3333 = \mathbf{10.3333 \ lb/in.^2}$$

9. What law allows a small force, moving a long distance, to move a large load a short distance?
   Answer: The law of conservation of energy, or energy (work) out equals energy (work) in.

11. How many inches of mercury (in. Hg) are there in one lb/in.^{2}?

   Answer: $\dfrac{29.92 \text{ in. Hg}}{14.7 \text{ lb/in.}^2}$ = **2.03537415 in. Hg per lb/in.^{2}**

13. Vacuum is measured in what units?
   Answer: in. Hg or lb/in.^{2} (absolute)

15. How much work is required at the pump of Figure 1.6 to lift the load 10 inches?
   Answer: 10 in. × 100 1b = **1000 in.-lb**

Chapter Two

1. How far will a 5-in.^{2} cylinder move if it receives 30 in.^{3} of fluid?

   Answer: $\dfrac{\overset{6 \text{ in.}}{\cancel{30 \text{ in.}^3}}}{\cancel{5 \text{ in.}^2}}$ = **6 in.**

3. True or false?: When dividing common bases containing exponents, merely add the exponent of the divisor to the exponent of the dividend (number being divided).
   Answer: False. You must *subtract* the exponent of the divisor from the exponent of the dividend.

5. How fast will the cylinder of Figure 2.9 move the load if a flow of 300 in.^{3}/min (300 cubic inches per minute) is directed to the cylinder area of 5 in.^{2}?

   Answer: $\dfrac{\overset{60 \text{ in.}}{\cancel{300 \text{ in.}^3}}}{\text{min}} \times \dfrac{1}{\cancel{5 \text{ in.}^2}}$ = **60 in./min**

7. Extending speed = _____ in./min.

   Answer: $\dfrac{3 \text{ gal}}{\text{min}} \times \dfrac{\overset{\text{in.}}{231 \text{ in.}^3}}{\text{gal}} \times \dfrac{1}{4 \text{ in.}^2}$ × = **173.25 in./min**

9. Pressure required to lift the load = _____ lb/in.^{2}.

   Answer: $\dfrac{\overset{250}{\cancel{1000} \text{ lb}}}{\cancel{4} \text{ in.}^2}$ = **250 lb/in.^{2}**

11. Flow from the rod end (annulus area) of the cylinder while extending = _____ gal/min.
   Answer: Flow from the rod end of a 2:1 cylinder is half the flow to the cap end; therefore:

$$\frac{3 \text{ gal}}{\text{min}} \times \frac{1}{2} = \textbf{1.5 gal/min}$$

13. Extending speed = _____ cm/min.

Answer: $\frac{3 \cancel{L}}{\text{min}} \times \frac{\overset{10 \text{ cm}}{\cancel{1000} \text{ cm}^3}}{\cancel{L}} \times \frac{1}{\underset{}{\cancel{100} \text{ cm}^2}} = \textbf{30 cm/min}$

15. Pressure required to lift the load = _____ kPa.

Answer: Pressure $= \frac{4000 \text{ N}}{\cancel{100} \text{ cm}^2} \times \frac{\overset{100}{\cancel{10,000} \text{ cm}^2}}{\text{m}^2}$

$= 400,000 \text{ N/m}^2$

$= 400,000 \text{ Pa} = \textbf{400 kPa}$

17. Flow from the rod end (annulus area) of the cylinder while extending = _____ L/min.
   Answer: The flow from the rod end of a 2:1 area cylinder while extending is one-half the cap-end flow; therefore, 3 L/min × 1/2 = **1.5 L/min.**

19. Calculate the (U.S.) horsepower output of a pump that supplies 5 gal/min flow at a pressure of 1500 lb/in.^{2}.

Answer: hp $= \dfrac{\text{psi} \times \text{gal/min}}{1714} = \dfrac{1500 \times 5}{1714} = \textbf{4.376 hp}$

21. Calculate the SI power output of a pump in kilowatts if the pressure is 100 bars and the flow is 12 liters per minute.

Answer: kW $= \dfrac{12 \text{ L/min} \times 100 \text{ bars}}{600} = \textbf{2 kW}$

Chapter Three

   1. True or False?: Hydraulic cylinders are linear actuators.
   Answer: True.
   3. True or False?: Ram-type cylinders are single acting.
   Answer: True.
   5. True or False?: Single-acting cylinders exert force in only one direction.
   Answer: True.
   7. True or False?: The extending and retracting speeds of a differential cylinder are equal.
   Answer: False.
   9. True or False?: A fixed flow to either end of a differential cylinder will produce equal speeds.
   Answer: False.

11. True or False?: The cylinder piston metal ring seal provides for better servo accuracy.
    Answer: True.

13. True or False?: Fixed-displacement hydraulic motors produce speed proportional to pressure.
    Answer: False.

15. True or False?: In a hydraulic gear motor, torque is the result of the hydraulic pressure acting against the area of two teeth.
    Answer: False. The forward torque on two teeth minus the reverse torque on one tooth equals the forward torque on one tooth.

17. True or False?: The radial load on the shaft bearing of the unbalanced vane motor is large.
    Answer: True.

19. True or False?: Piston motors have the advantage of lower internal leakage than any other type of motor.
    Answer: True.

21. True or False?: The in-line piston motor has less torsional efficiency than the bent-axis type.
    Answer: True.

23. How fast will a 3¼-in. cylinder with a heavy-duty rod retract when supplied with a fixed flow of 10 gal/min? (*Hint:* use 5.154 in.² as the annulus area of the 3¼-in. cylinder.)

    Answer: $\dfrac{10 \text{ gal}}{\text{min}} \times \dfrac{231 \text{ in.}^3}{\text{gal}} \times \dfrac{1}{5.154 \text{ in.}^2} = \textbf{448 in./min}$

25. How fast will the cylinder of Problem 24 retract if its annular area is half that of its cap end and is supplied by the same 20 L/min flow?
    Answer: The 2:1 cylinder will retract twice as fast as it extends, since the flow is constant.

27. What is the speed of a hydraulic motor with a displacement of 4 in.³/rev when supplied with a flow of 20 gal/min? (*Hint:* We want revolutions on top in the answer, so we start the solution with rev/4 in.³.)

    Answer: $\dfrac{\text{rev}}{4 \text{ in.}^3} \times \dfrac{231 \text{ in.}^3}{\text{gal}} \times \dfrac{20 \text{ gal}}{\text{min}} = \textbf{1155 rev/min}$

29. What is the torque developed by a 3 in.³/rev hydraulic motor when it has a differential pressure of 3000 lb/in.²?

    Answer: $\dfrac{3000 \text{ lb}}{\text{in.}^2} \times \dfrac{3 \text{ in.}^3}{\text{rev}} \times \dfrac{\text{rev}}{6.28 \text{ rad}} = 1433 \text{ (lb-in.)/rad}$

    $= \textbf{1433 lb-in. torque}$

Chapter Four

1. True or False?: The check valve is a two-way valve.
   Answer: False. It is a one-way control valve.

3. True or False?: The ball-spring check valve has a better seating arrangement than the poppet type.
   Answer: False.
5. True or False?: There are two active ports in a two-way valve.
   Answer: True.
7. How long will it take the accumulator of Figure 4.1 to charge to its rated capacity of 1600 cm^3 if the pump delivers 4 L/min? (*Hint:* We want minutes as the answer, so we turn 4 L/min upside down and start the solution with min/4 L. Now all that is necessary is to cancel liters and then cancel cm^3, and all we have left is minutes.)

Answer: $\dfrac{min}{4\ \cancel{L}} \times \dfrac{\cancel{L}}{1000\ \cancel{cm^3}} \times 1600\ \cancel{cm^3} = \textbf{0.4 min}$

9. How long will it take the PO check valve of Figure 4.5a to open fully, for reverse flow, if the 1-cm-diameter pilot piston travels 1.5 cm and the pilot flow is limited to 3 L/min?
   Preliminary Calculation: The area of the 1-cm-diameter piston is

$$\frac{\pi\ D^2}{4} = \frac{3.1416 \times 1^2}{4} = 0.7854\ cm^2$$

Hint: Again we want minutes as the answer, so we start with min/3 L then cancel out all other dimensions.

Answer: $\dfrac{min}{3\ \cancel{L}}\ \dfrac{\cancel{L}}{1000\ \cancel{cm^3}} \times 0.7854\ \cancel{cm^2} \times 1.5\ \cancel{cm} = \textbf{0.000392 min}$

$$0.0003927\ \cancel{min} \times \frac{60\ \cancel{sec}}{\cancel{min}} \times \frac{1000\ msec}{\cancel{sec}} = \textbf{23.562 msec}$$

Chapter Five

1A. What is the cracking pressure of the relief valve of Figure 5.1 if the area of the poppet piston is 0.5 in.^{2} and the spring is adjusted to a force of 35 lb?

Answer: $\dfrac{35\ lb}{0.5\ in.^2} = \textbf{70 lb/in.}^2$

1B. What is the cracking pressure of the relief valve of Figure 5.1 if the area of the poppet piston is 1 cm^2 and the spring is adjusted to a force of 50 kg? (*Hint:* 50 kg × (9.80665 N)/kg = 490.33 N; see Table 2.3.)

Answer: $\dfrac{490.33\ N}{1\ \cancel{cm^2}} \times \dfrac{10,000\ \cancel{cm^2}}{m^2} = \textbf{4,903,000 N/m}^2$

$$= \textbf{4,903,000 Pa}$$
$$= \textbf{4903 kPa}$$

3. What is the percent regulation of the valve of Figure 5.1 if the cracking pressure is 60 lb/in.^{2} and the pressure at full flow is 80 lb/in.^{2}? [*Hint:* Percent regulation is equal to the change divided by the original setting. Or (80 − 60)/60 = 20/60 = 0.33 × 100 =

33%. Now, redo Problem 3 using a 30 lb/in.^{2} pressure override with the same valve setting (60 lb/in.^{2}).]

Answer: Percent regulation = $30/60 \times 100 = $ **50%**

5A. How long will it take cylinder 1 of Problem 4A to retract fully if the rod diameter is 2 in.? [*Hint:* Rod area = $3.1416 \times (2 \text{ in.})^2/4$.]

$$\textit{Preliminary Calculation: Cap area} = \frac{3.1416 \times (3.25)^2}{4} = 8.3 \text{ in.}^2$$

$$\text{Rod area} = \frac{3.1416 \times (2)^2}{4} = 3.1416 \text{ in.}^2$$

$$\text{Annulus area} = 8.3 \text{ in.}^2 - 3.1416 \text{ in.}^2 = 5.1584 \text{ in.}^2$$

Hint: Again, we put the minutes on top and keep them there while canceling everything else.

Answer: $\dfrac{\text{min}}{3.5 \text{ gal}} \times \dfrac{\text{gal}}{231 \text{ in.}^3} \times 5.1584 \text{ in.}^2 \times 12 \text{ in.} = $ **0.0766 min**

$$0.0766 \text{ min} \times \frac{60 \text{ sec}}{\text{min}} = \textbf{4.2 sec}$$

5B. How long will it take cylinder 1 of Problem 4B to retract fully if the rod diameter is 5 cm? [*Hint:* Rod area = $3.1416 \times (5 \text{ cm})^2/4$.]

$$\textit{Preliminary Calculation: Cap area} = \frac{3.1416 \times (8 \text{ cm})^2}{4} = 50 \text{ cm}^2$$

$$\text{Rod area} = \frac{3.1416 \times (5 \text{ cm})^2}{4} = 19.635 \text{ cm}^2$$

$$\text{Annulus area} = 50 \text{ cm}^2 - 19.635 \text{ cm}^2 = 30.365 \text{ cm}^2$$

Hint: Again we invert the pump flow to put minutes on top, while canceling all other dimensions.

Answer: $\dfrac{\text{min}}{10 \text{ L}} \times \dfrac{\text{L}}{1000 \text{ cm}^3} \times 30.365 \text{ cm}^2 \times 30 \text{ cm} = $ **0.091 min**

$$0.091 \text{ min} \times \frac{60 \text{ sec}}{\text{min}} = \textbf{5.4657 sec}$$

7A. What will be the lifting speed (in./min) of the weight in Figure 5.6 if the cylinder has a 4-in. bore, a 2-in. rod, and the pump flow is 10 gal/min?

$$\textit{Preliminary Calculation: Cap area} = \frac{3.1416 \times 4 \times A}{A} = 12.5664 \text{ in.}^2$$

$$\text{Rod area} = \frac{3.1416 \times 2 \times 2}{4} = 3.1416 \text{ in.}^2$$

$$\text{Annulus area} = 12.5664 \text{ in.}^2 - 3.1416 \text{ in.}^2 = 9.4248 \text{ in.}^2$$

Answer: Lifting speed $= \dfrac{10 \text{ gal}}{\text{min}} \times \dfrac{231 \text{ in.}^3}{\text{gal}} \times \dfrac{1 \text{ in.}}{9.4248 \text{ in.}^2} = \mathbf{245 \text{ in./min}}$

7B. Rework Problem 7A using the following dimensions: bore diameter of cylinder = 10 cm; rod diameter of cylinder = 5 cm; and pump flow = 40 L/min.
Answer: Show all work.

9. True or False?: A two-stage relief valve has better pressure regulation than a single-stage type.
Answer: True.

11. True or False?: The spring-size problem of a single-stage relief valve is solved in the R-type design.
Answer: True.

13. True or False?: The R-type valve is a two-stage balanced-piston relief valve.
Answer: False.

15. True or False?: The bottom cover of the R-type sequence valve is arranged so as to allow control pressure to be taken from the primary system line.
Answer: True.

17. True or False?: A motor system requires the same counterbalance valve as a cylinder system.
Answer: False.

19. True or False?: The pressure override of a two-stage relief valve is small compared to that of a single-stage valve.
Answer: True.

21. True or False?: A closed switch is analogous to a closed valve.
Answer: False.

Chapter Six

1. True or False?: In most systems, flow is controlled by the variable restriction of a valve.
Answer: True.

3. True or False?: In the electrical analogy of hydraulics, the restriction is like a Zener diode.
Answer: False. The restriction is like a resistor.

5. True or False?: When resistors are connected in series, all voltages are the same.
Answer: False.

7. True or False?: The bleed-off flow control is the most power efficient of the three flow controls when used with a fixed-pressure pump.
Answer: False.

9. True or False?: The manually operated flow valve is not referred to as a proportional valve because it is not electrically operated.
Answer: True.

11. True or False?: Cylinders can only be used in systems with fixed-flow power supplies.
Answer: False. Cylinders may also be used with fixed-pressure systems.

13. From Figure 6.6b, what is the flow from the cap end of the cylinder while retracting?

Answer: The cap-end flow is twice the rod-end flow because the cap-end area is twice the annulus area. Therefore, the cap-end flow = 2 × 800 in.^{3}/min = **1600 in.^{3}/min.**

15. From Figure 6.7b, what is the flow from the cap end of the cylinder (F_c) during retraction?

Answer: As in Problem 13, the cap-end flow is twice the annulus flow since the cap-end area is twice that of the annulus. Or, the flow from the cap end is

$$2 \times 400 \text{ in.}^3/\text{min} = \textbf{800 in.}^3\textbf{/min}$$

17. From Figure 6-9b, what would be the flow through the 55-Ω resistor if the ac input to the primary of the transformer were increased to 120 V?

Answer: $I = E/R = 60/55 = \textbf{1.09 A}$

19. From Figure 6.10, recalculate the extending speed of the cylinder when the rod area is increased to 3 in.^{2} and the retracting speed is adjusted to 100 in./min.

Answer: $\text{Speed} = \sqrt{\dfrac{4}{1}} \times 100 \text{ in./min}$

$$= 2 \times 100 \text{ in./min} = \textbf{200 in./min}$$

21. From Figure 6.15, how much flow will be bypassed to tank by the compensator spool if the throttle is set at 7 gal/min? Answers: (a) in gal/min, (b) in in.^{3}/min, (c) in cm/min.

Answer: (a) 10 gal/min − 7 gal/min = **3 gal/min**

(b) $\dfrac{3 \text{ gal}}{\text{min}} \times \dfrac{231 \text{ in.}^3}{\text{gal}} = \textbf{693 in.}^3\textbf{/min}$

(c) $\dfrac{3 \text{ gal}}{\text{min}} \times \dfrac{3.7854 \text{ L}}{\text{gal}} \times \dfrac{1000 \text{ cm}^3}{\text{L}} = \textbf{11,356.2 cm}^3\textbf{/min}$

Chapter Seven

1. What will be the inlet pressure of the pump in Figure 7.1 if the pump handle is retracted 0.6 in. from the 2-in. stop prior to the start of flow? (*Hint:* Use Boyle's law.)

Boyle's law states $\dfrac{P_1}{P_2} = \dfrac{V_2}{V_1}$ where $P_1 = 14.5$ psia (the original pressure) and

$V_1 = 2$ in.^{3} (1 in.^{2} × 2 in.) and $V_2 = 2.6$ in.^{3}.

Answer: $\dfrac{14.5}{P_2} = \dfrac{2.6}{2}$

$2.6\, P_2 = 14.5 \times 2$ (cross multiplication)

$P_2 = \dfrac{14.5 \times 2}{2.6}$

$P_2 = 11.15$ psia Pump inlet pressure (absolute)

$= 11.15 - 14.5 = -3.35$ psig Pump inlet pressure (gauge)

$$= \frac{3.35}{14.5} \times 29.95 \text{ in. Hg} = \frac{6.9 \text{ in. Hg}}{\text{pump inlet (vacuum)}}$$

3. What mechanical horsepower will be required of the motor driving the pump in Problem 2, if the overall efficiency of the pump is 78%?

   Answer: The output horsepower of the pump of Problem 2 is

$$\frac{5 \text{ gal/min} \times 1714 \text{ psi}}{1714} = 5 \text{ hp}$$

   The input horsepower required of the 78% efficient pump is

$$\frac{5 \text{ hp}}{0.78} = \textbf{6.41 hp}$$

5. What is the input horsepower to the pump in Figure 7.5 if the speed is 1800 rev/min and the torque to the pump shaft is 250 lb-in.?

   Answer: Pump input hp $= \dfrac{1800 \text{ rev/min} \times 250 \text{ lb-in.}}{63,025}$

$$= \textbf{7.14 hp}$$

7. What is the output power in kW of the pump in Figure 7.5 if it delivers 28 L/min at a pressure of 240 bars?

   Answer: Output power $= \dfrac{28 \text{ L/min} \times 240 \text{ bars}}{600} = \textbf{11.2 kW}$

9. What is the overall efficiency of the pump in Figure 7.5 if it has the input horsepower of Problem 5 and the output horsepower of Problem 8?

   Answer: Overall efficiency $= \dfrac{\text{Output hp} \times 100}{\text{Input hp}}$

$$= \frac{5.83}{7.14} \times 100 = \textbf{81.7\%}$$

11. Electric drive motor overall efficiency is:

    Preliminary Calculation: Motor input hp $= 1.12 \text{ k\cancel{W}} \times \dfrac{\text{hp}}{0.746 \text{ k\cancel{W}}} = 1.5 \text{ hp}$

    Answer: Motor overall efficiency $= \dfrac{\text{Motor output hp} \times 100}{\text{Motor input hp}}$

$$\frac{1.3}{1.5} \times 100 = \textbf{87\%}$$

13. System overall efficiency is:

    Preliminary Calculation: Pump output hp $= \dfrac{1.75 \text{ gal/min} \times 1000 \text{ psi}}{1714}$

$$= \textbf{1.02 hp}$$

Answer: System overall efficiency $= \dfrac{\text{Pump output hp}}{\text{Motor input hp}} \times 100$

$$= \frac{1.02}{1.5*} \times 100 = \mathbf{68\%}$$

Note: See the preliminary calculation from Problem 11.

15. True or False?: Flow is the only parameter of a pump that can be fixed.
    Answer: False. Pressure can also be fixed, as in a pressure-compensated pump.

17. True or False?: Fixed-flow positive displacement pumps have fixed flow that does not change with speed.
    Answer: False. Flow will change with pump speed.

19. True or False?: In the SI system of standards, there are 100 bars in an atmosphere.
    Answer: False. One bar = one atmosphere.

21. True or False?: To prevent excessive pump cavitation, the pump inlet should be limited to an absolute pressure of not less than 12 lb/in.^{2} (a).
    Answer: True.

23. True or False?: Variable displacement pump means that the volume of fluid per revolution (in.^{3}/rev) is adjustable.
    Answer: True.

25. True or False?: There must be some restriction in the load in order for a pump to develop pressure.
    Answer: Absolutely true.

27. True or False?: The external gear pump has an unbalanced side load on its bearings.
    Answer: True.

29. True or False?: The gerotor pump has a fixed displacement and a balanced bearing load.
    Answer: False. The bearing load is unbalanced.

31. True or False?: Reciprocating pistons make poor pumps.
    Answer: False. Reciprocating pistons make good pumps.

33. True or False?: The pressure-compensated pump has an imbalance of forces acting on its yoke while holding a constant pressure.
    Answer: False. The yoke forces are balanced or the yoke would not have stopped moving.

Chapter Eight

1. How fast will the cylinder of Figure 8.1b extend if the pump delivers 10 gal/min?

    Answer: $\dfrac{10 \text{ gal}}{\text{min}} \times \dfrac{231 \overset{\text{in.}}{\text{in.}^3}}{\text{gal}} \times \dfrac{1}{10 \text{ in.}^2} = \mathbf{231 \text{ in./min}}$

3. How fast will the cylinder of Figure 8.11 extend if its cap-end area is 10 cm^2 and the pump delivers 10 L/min?

    Answer: $\dfrac{10 \text{ L}}{\text{min}} \times \dfrac{1000 \overset{\text{cm}}{\text{cm}^3}}{L} \times \dfrac{1}{10 \text{ cm}^2} = \mathbf{1000 \text{ cm/min}}$

$$\text{Or } \frac{\overset{10}{\cancel{1000}\,\cancel{cm}}}{min} \times \frac{m}{\cancel{100}\,\cancel{cm}} = \textbf{10 m/min}$$

$$\text{Or } \frac{1000\ cm}{\cancel{min}} \times \frac{\cancel{min}}{60\ sec} = \textbf{16.67 cm/sec}$$

Note: Since $\dfrac{m}{100\ cm} = 1$, multiplying by this term will not change the absolute value of the expression, only the dimension.

5. What would be the minimum storage capacity of the tank of Figure 8.1 if there were five cylinders in the circuit with the same rod and cap areas, but each had a stroke length of 25 in.? (*Note:* You must store enough fluid volume in the tank to equal the volume of all the cylinder rods of the system. Otherwise you would run out of fluid before the last cylinder could extend.)

   Answer: The volume of one cylinder rod is:

   $$5\ in.^2 \times 25\ in. = 125\ in.^3$$

   The total rod volume is:

   $$5 \times 125\ in.^3 = \textbf{625 in.}^3 \text{ (the minimum capacity of the tank)}$$

7. What would be the extending speed of the cylinder of Figure 8.3 if the retracting speed is 250 in./min and all other parameters are the same? (*Note:* The extending speed of an unloaded cylinder with a pressure-compensated pump and a noncompensated flow control is approximately equal to the retracting speed times the square root of the ratio of the cap-end area to the rod-end annulus area.)

   Answer: Extending speed $= 250 \times \sqrt{2/1}$
   $$= 250 \times \sqrt{2}$$
   $$= 250 \times 1.414 = \textbf{353.5 in./min}$$

9. What would be the extending speed of the cylinder of Figure 8.3 if the retracting speed is 2500 cm/min, while the ΔP retracting across the flow valve $= 1,000,000$ N/m^2 (1000 kPa) and the ΔP extending $= 2,000,000$ N/m^2 (2000 kPa)? *Note:* Speed of the cylinder is directly proportional to the square root of the ratio of the ΔP extending to ΔP retracting.

   Answer: Extending speed $= 2500$ cm/min $\times \sqrt{2}$
   $$= 2500\ cm/min \times 1.414 = \textbf{3535 cm/min}$$
   $$\text{or } \frac{3535\ \cancel{cm}}{min} \times \frac{m}{100\ \cancel{cm}} = \textbf{35.35 m/min}$$

11. True or False?: At standby, the pump delivery of Figure 8.1a is being dumped to tank through the tandem-centered three-position four-way valve (V_2).

    Answer: True.

13. True or False?: In the circuits of this chapter, the pump and electric motor are shown on the right-hand side of the drawings.

    Answer: False. They are shown on the left-hand side.

15. True or False?: The tandem-centered valve is sometimes used in nontandem circuits.
Answer: True.

17. True or False?: The tank must have the capacity to store all cylinder rods of the system.
Answer: True.

19. True or False?: The result of a 2:1 area ratio regenerative cylinder system with a fixed flow is equal extension and retraction speeds.
Answer: True.

21. True or False?: To prevent coil burnout, design the electrical circuit so that both solenoids of a single valve are energized at the same time.
Answer: False. This would ensure a coil burnout.

23. True or False?: Check valve V_1 of Figure 8.7 is used to discharge the accumulator.
Answer: False. It is used to prevent the accumulator from discharging back through the pump.

25. True or False?: The pressure switch of Figure 8.8 both vents and devents the relief valve.
Answer: True.

27. True or False?: In Figure 8.10, the accumulator is discharged automatically when solenoid A of V_2 is energized.
Answer: False. This happens when solenoid A is deenergized.

29. True or False?: The motor of Figure 8.12 will run as a pump during deceleration because of the load inertia.
Answer: True.

31. True or False?: The 50 lb/in.^{2} check valve in Figure 8.14 assures that pilot pressure is always available for V_2, even when it is in center position.
Answer: True.

33. True or False?: The small circles to the right of the electrical drawing are the relay coils.
Answer: True.

35. True or False?: Limit switches in electrical drawings are always shown in their open position.
Answer: False. They are shown in their held position.

37. True or False?: On a sequence diagram, 1CR * is a symbol meaning that "the number one control relay was deenergized at that point."
Answer: True.

39. True or False?: The check valve circuit of Figure 8.22 is equivalent to the four diodes used in a bridge rectifier circuit.
Answer: True. They ensure that flow through the relief valve is always in the same direction regardless of which side of the system has high pressure. The bridge rectifier, in like manner, ensures that current flows in the same direction through the load (dc) even though the input is ac.

Chapter Nine

1. True or False?: Energy is consumed while going through a system.
Answer: False. Energy *out* equals energy *in;* some of the output energy, however, may be in the form of heat due to the inefficiencies of the system.

3. True or False?: The parameters shown across the bottom of Figure 9.1 have their influence from left to right.
   Answer: False. Right to left.

5. True or False?: To regulate load speed completely, we must maintain constant speed under all changing load conditions.
   Answer: True.

7. True or False?: The best way to regulate speed is to use *load compensation.*
   Answer: False. The best way is to use servo feedback.

9. True or False?: The best method of servo control is to feed back a *hydraulic* signal to the *electrical* part of the system.
   Answer: False. It is better to feed back the electrical signal to the hydraulic part of the system.

11. True or False?: If the load of a velocity servo slows down, the tachometer will generate less voltage.
    Answer: True.

13. True or False?: Low oil compressibility is a key to the response of the electrohydraulic servo system.
    Answer: True.

15. True or False?: Pressure-compensated pumps are not well suited for servo power supplies.
    Answer: False.

17. True or False?: The piston motor performs poorly in a servo system because it has a high breakaway pressure.
    Answer: False. Although the piston motor has a larger breakaway than the vane motor, the low leakage of the piston motor along with the high pressure gain of the servo valve *make the piston motor the best servo motor.*

19. True or False?: The stylus of a tracer valve may be pushed into the valve by the force of your little finger.
    Answer: True. This is the same as the low force needed to move the steering wheel of an auto while the power steering system is actually steering the car.

21. Assuming 100% efficiency throughout the system of Figure 9.1, what is the flow through the hydraulic motor if the hydraulic pressure is 1714 lb/in.^{2} and the electric input is 746 V at 10 A? (*Note:* One electric horsepower = 746 W, and volts × amperes = watts.)

    Answer: Input electric hp $= 10 \times 746\,\text{W} \times \dfrac{1\ \text{hp}}{746\,\text{W}} = \textbf{10 hp}$

    The output hydraulic hp must also = 10 hp (100% eff.)

    $$\text{hp} = \frac{\text{psi} \times \text{gal/min}}{1714} \quad \text{(Reference Chapter 2)}$$

    $$\text{gal/min} = \frac{1714\ \text{hp}}{\text{psi}} \quad \text{(Draw from your algebra background)}$$

    $$= \frac{1714 \times 10}{1714} = \textbf{10 gal/min}$$

23. What will be the acceleration (a = in./sec^2) of a load if the force is (f = 1000 lb) and the mass is (m = 100 lb-sec^2/in.)? Formula: ($a = \dfrac{f}{m}$).

Answer: $a = \dfrac{f}{m}$

$$= \dfrac{\overset{10}{\cancel{1000\ \text{lb}}}}{\cancel{100\ \text{lb}}\ \text{sec}^2/\text{in.}^*} = \textbf{10 in./sec}^2 \text{ or 10 in. per second per second}$$

Note: "in." is moved from the bottom of the bottom of the equation to the top. (Again, draw from your algebra background.)

25. What is the flow through a meter motor rotating at 1000 rev/min if it has a displacement of 2.31 in.^{3}/rev? (*Note:* A meter motor is a motor calibrated in gal/min.)

Answer: $\dfrac{2.31\ \text{in.}^3}{\cancel{\text{rev}}} \times \dfrac{1000\ \cancel{\text{rev}}}{\text{min}} = 2310\ \text{in.}^3/\text{min}$

$$\text{or} = \dfrac{\overset{10}{\cancel{2310\ \text{in.}^3}}}{\text{min}} \times \dfrac{\text{gal}}{\cancel{231\ \text{in.}^3}} = \textbf{10 gal/min}$$

27. True or False?: When the spool of the four-way servo valve is at the center, a small flow is metered from the pressure port, past the A and B cylinder ports to tank.
    Answer: True.

29. True or False?: Dither keeps the spool of a servo valve active through center.
    Answer: True. Properly adjusted dither keeps everything in the loop moving, except the load cylinder or motor.

31. True or False?: The no-amplifier servo system has excellent positional accuracy.
    Answer: False. Excellent accuracy requires more electrical gain.

33. True or False?: The voltage gain of an operational amplifier is equal to R_o/R_i.
    Answer: True. Gain = output resistance/input resistance.

35. True or False?: A servo system automatically corrects its output to correspond to its input or command.
    Answer: True.

37. True or False?: The "cookbook" approach to servo system analysis is based on much testing and experience.
    Answer: True.

39. True or False?: The tracking error of a servo cylinder is inversely proportional to the cylinder speed.
    Answer: False. Tracking error is directly proportional to cylinder speed.

41. True or False?: The best response of a velocity servo system is usually obtained with a linear amplifier.
    Answer: True. The system does not have to wait for the integrating amplifier.

43. True or False?: The first step in trimming a system is to check for hydraulic leaks and proper phasing.
    Answer: True.

45. True or False?: Preliminary flushing is strongly recommended to ensure the removal of any contaminants from the system before installing the servo valve.
    Answer: True.

47. What would be the power rating of the pots of Figure 9.11 if the excitation voltage were raised to 150 V?

Answer: Each pot must be able to take its excitation current of $I = E/R = 150/10,000 =$ **0.015 A,** plus the current input to the amplifier: $I = E/R = 150/100,000 =$ **0.015 A,** where the input resistance of the amplifier is assumed to be 10 times the pot resistance. Therefore, the wattage rating must be more than:

$$W = E \times I = 150 \times 0.0165 = \textbf{2.475 W}$$

So a **2.5-W** pot would be fine.

49. What is the valve-cylinder resonance of the system of Figure 9.20 with the following assumptions?:

$$\text{Cylinder annulus area} = 7.5 \text{ in.}^2$$
$$\text{Volume of fluid under compression} = 90 \text{ in.}^3$$
$$\text{Load weight} = 4000 \text{ lb}$$
$$\text{Compressibility of oil } (E) = 5 \times 10^{-6} \text{ in.}^2/\text{lb}$$

Answer: $\omega = \sqrt{\dfrac{2A^2}{VEM}}$

where

ω = resonant frequency (1/sec) or rad/sec

A = annular area of cylinder piston (in.^{2})

V = one-half the total volume of oil in the cylinder plus one-half the volume of oil in both lines (in.^{3})

E = compressibility of oil (5×10^{-6} in.^{2}/lb) (the reciprocal of bulk modules of oil in lb/in.^{2})

M = load mass = $\dfrac{\text{load wt (lb)}}{386 \text{ (in./sec}^2)}$ = lb-sec^2/in.

Note: To obtain the resonant frequency in hertz, divide the answer of rad/sec or 1/sec by 2π or 6.28.

Note: Acceleration of a free-falling body; taken from the physics formula $M = F/a$.

Preliminary Calculation: Load mass = $\dfrac{4000 \text{ lb}}{386 \text{ (in./sec}^2)}$ = **10.36 lb-sec^2/in.**

Answer: $\omega = \sqrt{\dfrac{2 \times (7.5 \text{ in.}^2)^2}{45 \text{ in.}^3 \times 5 \times 10^{-6} \text{ in.}^2/\text{lb} \times 10.36 \text{ lb-sec}^2/\text{in.}}}$

$= \sqrt{48262/\text{sec}^2}$

$= \textbf{220/sec}$ or **220 rad/sec** or $\dfrac{220 \text{ rad}}{\text{sec}} \dfrac{\text{cycle}}{2\pi \text{ rad}} = \textbf{35 cycles/sec}$

51. What is the recommended amplifier gain (G_a) setting of the system of Problem 49 with the following additional assumptions:

$$G_{sv} = 0.1 \ \frac{\text{in.}^3/\text{sec}}{\text{mA}}$$

$$H_{fb} = 5 \ \text{V/in.}$$

$$G_{cyl} = 1/7.5 \ \text{in.}^2$$

Answer: To ensure system stability, the velocity constant or loop gain (K_V) should be limited to one-third that of the lowest resonant frequency in rad/sec:

$$\text{Maximum loop gain} = \frac{220/\text{sec}*}{3} = 73.3/\text{sec}$$

Note: From Problem 49.

Answer: Now, K_V is the product of all the gains in the loop, or 73.3/sec = $G_a \times G_{sv} \times G_{cyl} \times H_{fb}$

$$\text{or } G_a = \frac{73.3/\text{sec}}{G_{sv} \times G_{cyl} \times H_{fb}}$$

$$\text{and } G_a = \frac{73.3/\text{sec}}{0.1 \ \text{in.}^3/\text{mA} \times 1/7.5 \ \text{in.}^2 \times 5 \ \text{V/in.}} = \textbf{1100 mA/V}$$

53. Using the instructions of Application Note 1 of the cookbook approach to accuracy predictions (Section 9.5), estimate the accuracy of a pot–pot system with the following parameters:

$$\text{Cylinder area} = 8 \ \text{in.}^2$$

$$\text{Cylinder length} = 20 \ \text{in.}$$

$$\text{Load weight} = 1000 \ \text{lb}$$

$$\text{Load force} = 4000 \ \text{lb}$$

$$\text{Supply pressure} = 1000 \ \text{lb/in.}^2$$

Answer: Error = 0.001 in. (basic)

+0.001 in. (additional 10-in. stroke; instruction 2)

Total **0.002 in.** × 1 (for a 3¼-in. bore)

55. Calculate the closed-loop droop of a linear velocity servo system when the following information is known: (1) the open-loop droop = 25 rev/min (with load); (2) the feedback tach voltage = 10 V, with (3) a command voltage = 15 V.

Answer: Closed-loop droop = $25 \ \text{rev/min} - \dfrac{(25 \ \text{rev/min} \times \text{feedback V})}{\text{command V}}$

$$= 25 \ \text{rev/min} - \frac{(25 \ \text{rev/min} \times 10 \ \cancel{V})}{15 \ \cancel{V}}$$

$$= \textbf{8.33 rev/min}$$

Chapter Ten

1. True or False?: The most limiting factor of the well-maintained system is the bearing life expectancy of its rotating parts.
   Answer: True.

3. True or False?: Most solenoid-operated valves may be manually operated.
   Answer: True.

5. True or False?: Excessive noise means excessive wear.
   Answer: True.

7. True or False?: Restriction is required to produce pressure.
   Answer: True.

9. True or False?: Excessive pump noise may be caused by air in the hydraulic fluid.
   Answer: True.

11. Why do we sometimes use a larger sized cylinder in a servo system than is actually needed to move the load?
    (A) We want to increase the load speed.
    (B) We want to lower the acceleration of the load.
    (C) We want to increase the positional accuracy.
    (D) We want to move the load more easily through the "hard spots."

    Answer: C.
    Calculate the following efficiencies for a hydraulic system with the inputs and outputs listed in the *Note.

13. What would be the maximum extending speed of a 2:1 area ratio servo cylinder, if its maximum retracting speed is 100 in./min?
    Answer: Extending speed $= 100 \text{ in./min} \times \sqrt{2}$
    $$= 100 \text{ in./min} \times 1.414$$
    $$= 141.4 \text{ in./min}$$

15. Electric drive motor overall efficiency.

    Preliminary Calculation: Input electric hp $= 1.5 \text{ kW} \times \dfrac{1 \text{ hp}}{0.745 \text{ kW}} = 2 \text{ hp}$

    Answer: Electric motor overall efficiency $= \dfrac{1.5 \text{ hp}}{2.0 \text{ hp}} \times 100 = \mathbf{75\%}$

17. System overall efficiency.

    Preliminary Calculation: Pump output hp $= \dfrac{2 \text{ gal/min} \times 1000 \text{ lb/in.}^2}{1714} = 1.17 \text{ hp}$

    Answer: System overall efficiency $= \dfrac{1.17 \text{ hp}}{2.00 \text{ hp*}} \times 100 = \mathbf{58.5\%}$

    Note: From the preliminary calculation of Problem 15. The inputs and outputs follow:

Electric drive motor: Input = 1.5 kW

Output = 1.5 hp

Hydraulic pump: Input = 1.5 hp

Output flow = 2 gal/min

Output pressure = 1000 lb/in.^{2}

Index

R